Jan Harmsen

Chemical Engineering: 25 Must-Know Process Design Principles

De Gruyter Graduate

Also of Interest

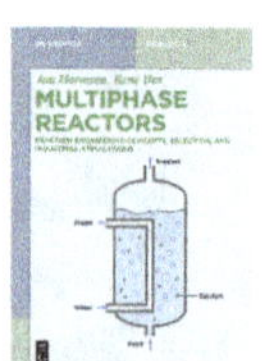

Multiphase Reactors.
Reaction Engineering Concepts, Selection, and Industrial Applications
Harmsen, Bos, 2023
ISBN 978-3-11-071376-3, e-ISBN (PDF) 978-3-11-071377-0,
e-ISBN (EPUB) 978-3-11-071384-8

Product and Process Design.
Driving Sustainable Innovation
Harmsen, de Haan, Swinkels, 2024
ISBN 978-3-11-078206-6, e-ISBN (PDF) 978-3-11-078212-7,
e-ISBN (EPUB) 978-3-11-078226-4

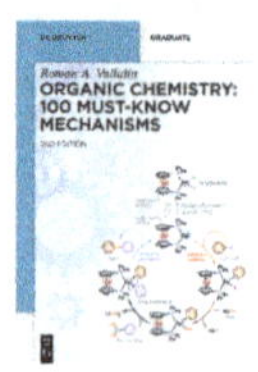

Organic Chemistry: 100 Must-Know Mechanisms
Valiulin, 2023
ISBN 978-3-11-078682-8, e-ISBN (PDF) 978-3-11-078683-5,
e-ISBN (EPUB) 978-3-11-078701-6

Organic Chemistry: 25 Must-Know Classes of Organic Compounds
Elzagheid, 2024
ISBN 978-3-11-138199-2, e-ISBN (PDF) 978-3-11-138275-3,
e-ISBN (EPUB) 978-3-11-138330-9

Toxicity: 77 Must-Know Predictions of Organic Compounds.
Including Ionic Liquids
Keshavarz, 2023
ISBN 978-3-11-118912-3, e-ISBN (PDF) 978-3-11-118967-3,
e-ISBN (EPUB) 978-3-11-119092-1

Jan Harmsen

Chemical Engineering: 25 Must-Know Process Design Principles

DE GRUYTER

Author
Ir. Jan Harmsen
Harmsen Consultancy B. V.
Hoofdweg Zuid 18
2912 ED Nieuwerkerk aan den IJssel
The Netherlands

ISBN 978-3-11-120323-2
ISBN 978-3-11-120325-6 (PDF)
ISBN 978-3-11-120401-7 (E-PUB)
DOI https://doi.org/10.1515/9783111203256

Library of Congress Control Number: 2026937926

Bibliographic information published by the Deutsche Nationalbibliothek
The Deutsche Nationalbibliothek lists this publication in the Deutsche Nationalbibliografie;
detailed bibliographic data are available on the Internet at https://dnb.dnb.de.

De Gruyter and Walter de Gruyter GmbH are part of De Gruyter Brill.
www.degruyterbrill.com

Questions about General Product Safety Regulation: productsafety@degruyterbrill.com

Cover illustration: farakos / iStock / Getty Images Plus

Contents

Part I: **Process design**

Part II: **Embedded process design**

Part IV: **Unit operations design**

Part V: **Scale-up by design**

Preface

Karin Sora, director Science, Technology, Engineering and Mathematics (STEM), of De Gruyter, mentioned to me at the CHISA conference 2022 that the book, "A 100 you must know chemical synthesis methods," sells well, and that she would love to have similar book on chemical engineering. This idea crept into me and stayed there. So, at the end of the conference, I decided to propose to author this book and Karin accepted that immediately.

Writing this book often was a pleasure, but sometimes it was a struggle—what to put in and what leave out—keeping in mind that the content should be broadly applicable, and the method should be easy to use. This ultimately helped to decide what to leave out.

https://doi.org/10.1515/9783111203256-203

Acknowledgment

I thank the reviewers Johan Grievink, Professor André de Haan, and Professor Tony Kiss, all professors at Delft University of Technology in the field of chemical engineering and also experienced in process design in various process industries. They provided elaborate comments on content, structure, and scope, which greatly improved the book quality. It took me two more months to incorporate all their comments in the manuscript. This resulted in more comprehensive and systematic content. Their comments also provided a better book scope.

I thank Silvia Polverini, Energy and Resources Management Consultant at Schneider Electric Sustainability Business, Ana Carolina Alves, Process Engineer at Battolyser Systems BV, Ezgi Arslanturkoglu, Process Development Engineer at Shell Global Solutions International BV, Tamoor Mughal, Researcher at Shell, Federico Consonni, Process Development Engineer at Shell Global Solutions International BV, and EggXpert B.V. Company, for their contributions to Case 6.3 in Chapter 6.

https://doi.org/10.1515/9783111203256-204

Introduction

Book scope

New process designs are more needed than ever before, because of the large problems the world community faces. To solve these problems, the Sustainable Development Goals of the United Nations are stated. To reach those goals, new industrial processes will be needed.

This book provides generic design methods to generate these new processes for all kinds of process industry branches, such as biofuels, food, water, chemicals, mineral ore processing, and pharmaceuticals.

The design methods are suitable to quickly generate process concepts, industrial symbiosis, and circular economy options.

The methods include rapid evaluations for Safety, Health, Environmental, Economic, Technical feasibility, and Sustainability (SHEETS).

The book also provides generic scale-up methods for processes as well as for individual process steps.

As for process design and scale-up, communications between different disciplines and different stakeholders are important, and an extensive vocabulary list is provided.

The content is based on established textbook design knowledge but goes beyond this by adding industrial design practices obtained from three decades of industrial experience in process research, development, design, and operation. This industrial knowledge is made plausible where needed. The body of knowledge therefore goes beyond commonly acquired academic knowledge. The design items based on industrial experience are made plausible by reasoning.

The book will therefore be useful for chemical engineers in industry, contract research organizations, and universities, who want to make a process concept design for any process industry, such as biofuels, mineral ore processing, chemicals, pharma, or dairy. It also provides scale-up methods from the concept design to commercialization.

A main difference of this book, compared to other process design textbooks, is that it is a design aid which needs no computer design packages, but only a calculator. By this, it facilitates rapid concept designs.

Book content

The Chemical Engineering Dictionary states that Chemical Engineering is a branch of engineering that deals with design, construction, and the operation of processes that involve physical, chemical, and biological change for the conversion of raw materials into useful materials on an industrial scale [1]. This book is therefore about process concept design and about process scale-up methods to industrial scale.

https://doi.org/10.1515/9783111203256-001

This book focuses furthermore on the early innovation stages of projects, where ideas are generated, exploratory experiments have been executed, and preliminary designs are made. The question to be answered by those designs is: Can this new idea lead to commercial scale success?

The process concept design method facilitates answering this question. The method provides a workflow for making a process concept design in a few hours to a few days. It is based on heuristics and using algebraic expressions for sizing streams and unit operations.

To this end, this book treats basic design methods for integral process design, as well as for unit operations design. Process concept design methods include process synthesis, process intensification, process heat integration, industrial symbiosis, circular economy, and life-cycle analysis.

Unit operations design methods include chemical reaction engineering, distillation, liquid-liquid extraction, adsorption, mixing, heat transfer, and particle design.

With the resulting design, a rapid assessment can be made on health, safety, environment, economics technical feasibility, and sustainable development, with methods provided in this book.

If the design seems feasible and attractive, an experimental scale-up program can be started. To this end, scale-up methods for integrated process as well as for unit operations are provided.

A vocabulary chapter is added to clarify terms used.

Each chapter provides: (1) Basic and background information on the method; (2) A concept design method, which can be quickly applied requiring only a calculator; (3) A case in which the design method is applied.

The process concept design method chapters treat first selecting promising process function steps and their connection from feed streams to product outlet streams. Second, for each process function the most attractive unit operation is selected. Third, the unit operations are sized with around 50 % accuracy.

The unit operations design chapters and treat design methods for most common unit operations.

The scale-up design chapters treat selection of integral process design options and selection of unit operations in view of scale-up risk reduction.

Intended book users

The intended users of this book are from a variety backgrounds. Here are five types of intended users of this book.

The first intended users are chemical engineers in industry, who design new processes. This book will help them to quickly apply a process design method. It will also help them to check whether this missed concept design options. It will also facilitate a comprehensive evaluation of their design.

The second intended users are experienced chemical engineers in industry who teach young chemical engineers how to design processes. This book provides them with all relevant information and the cases in each chapter show them how the information is used.

The third intended users are engineers, other than chemical engineers, such as physical, mechanical, civil, electronical engineers, who want to know what chemical engineers do, because they interact with them, or who want to design a process. Because of their engineering education, they will easily understand how they can use this book for that purpose.

The fourth intended users of this book are industrial chemists, or biologists, who research and develop (bio)chemical processes. By reading this book, they will be able to enrich their process design with mass and atom balances and will be able to represent their process design to others more clearly.

The fifth intended users are industry managers who want to find ways to incorporate circular economy and industrial symbiosis methods to convert waste into useful products and to reduce their environmental impact.

The sixth intended users are academics who teach process and unit operation design, as part of chemical engineering education, and who want to add industrial viewpoints and industrial cases to their education.

Historic developments of chemical engineering leading to this book

The field of chemical engineering started in 1901 with the book by Davis, Handbook of Chemical Engineering [2]. It describes design methods based on general physical operation principles. These general operation principles were a little later named unit operations by Arthur D. Little [3]. The latter also established chemical engineering as an education program by his school of Chemical Engineering Practice at MIT. This then delivered trained chemical engineers to the industry [4]. The book "Principles of Chemical Engineering" by Walker et.al. first published in 1923 [5] helped to establish academic education in many universities worldwide. The first edition of Perry's Chemical Engineers's Handbook, published in 1934 [6] aided industrial designers to quickly find design information for each unit operation.

This new chemical engineering design method had the additional advantage that now a scale-up method could be defined in which the pilot plant was designed with the same unit operations as the envisaged commercial scale plant. This meant that no longer endless trial and error experiments were needed to develop a process, but that a commercial scale concept design as well as the pilot plant design were made with the same unit operations. The theoretical base of the unit operations, also in view of scale-up was further developed by introducing dimensionless numbers.

With the introduction of stage-gate innovation methods with the stages, ideation, concept, feasibility, development, Engineering Procurement and Construction (EPC) and commercial scale start-up the chemical engineering discipline knowledge could focus on the stages prior to the EPC stage. Academic education then limited itself to design methods for the concept stage.

Chemical reaction engineering became an established unit operation during the period 1950–1970 [7]. The word "Chemical" refers here to chemical reactions, so changes in the molecular composition of the substances are treated in the reactor. The theory can be applied to any industrial process design involving molecular changes, such as food, plastics, ceramics, minerals, oil refineries, pharmaceuticals, and chemicals.

Process intensified unit operations started in the 1970s and now process intensified principles to generate new unit operations, which have been defined [8].

A new way of designing intensified chemical processes based on functions started in the 1970s by Siirola [9]. This new way was applied successfully in industries for breakthrough industrial processes for chemicals as well as for biofuels processes [10].

By the end of the twentieth century, industrial symbiosis applications emerged, in which industrial plants were connected in industrial complexes and urban systems such as district heating and domestic wastewater treatment were connected with industrial plants [11].

These major chemical engineering design methods developed in the last 100 years are treated in this book. Its main purpose is a quick reference book to design an early concept process design, requiring little information and only a calculator. That early design can then be used to evaluate ideas on Health Safety, Environment, Economics, Technical feasibility, and Sustainability (SHEETS) [12].

Bibliography

[1] Schaschke C. A Dictionary of Chemical Engineering. Oxford: OUP; 2014 Jan 9.

[2] Davis GE. In: A Handbook of Chemical Engineering: Illustrated with Working Examples and Numerous Drawings from Actual Installations. Davis Bros.; 1904.

[3] Little AD, et al. Chemistry in history, Chemical Heritage Foundation, Retrieved 13 Nov. 2013. Resourced from https://wikipedia.org/w/index.php?title=Unit_operation&oldid=709344969.

[4] Peppas NA, editor. One Hundred Years of Chemical Engineering: From Lewis M. Norton (MIT 1888) to Present. Springer; 1989 Apr 30.

[5] Walker WH, Lewis WK, McAdams WH. Principles of Chemical Engineering. 1st ed. New York: McGraw-Hill; 1923.

[6] Perry JH. Perry's Chemical Engineers Handbook. 1st ed. New York: McGraw-Hill; 1934.

[7] Harmsen J, Bos R. Multiphase Reactors: Reaction Engineering Concepts, Selection, and Industrial Applications. De Gruyter; 2023 Apr 12.

[8] Stankiewicz A, Van Gerven T, Stefanidis G. The Fundamentals of Process Intensification. Wiley; 2019.

[9] Siirola JJ, Rudd DF. Computer-aided synthesis of chemical process designs. From reaction path data to the process task network. Industrial & Engineering Chemistry Fundamentals. 1971 Aug;10(3):353–62.

[10] Harmsen J, Verkerk M. Process Intensification: Breakthrough in Design, Industrial Innovation Practices, and Education. Walter de Gruyter GmbH & Co KG; 2020.

[11] Harmsen J, Powell JB. Sustainable development in the process industries. Hoboken, NJ: John Wiley & Sons; 2010.
[12] Harmsen J, De Haan AB, Swinkels PLJ. Product and Process Design: Driving Sustainable Innovation. Berlin: De Gruyter; 2024.

Part I: **Process design**

1 Process concept design based on conventional unit operations

1.1 Basics process concept design

Process concept design, also called process synthesis, is located in the concept innovation stage after the discovery stage and prior to the feasibility and development stages. Process concept design defines all process input and output streams in composition and size. It also defines all unit operations or process steps, to transform the input streams to the output streams. It furthermore defines all internal streams, including recycling streams inside the process design.

It may contain heat integration systems to reduce energy requirements, but in industry, heat integration is in general carried out as part of detailed engineering in the Engineering, Procurement, Construction (EPC) stage.

An alternative term for process concept design is process synthesis. In this chapter, the term process concept design is used for designing with conventional unit operations. The main reason for using this term is that process synthesis is reserved by some authors for process design based on functions and function integration. That design method, called process intensification, is treated in Chapter 2.

Total mass balance

The first basic element of process concept design is that the total mass input is equal to the total mass output. This steady-state averaged mass balance holds for long periods such as a year. Therefore, the mass balance of the process is expressed in mass per year, and the total mass input of the process is equal to the mass output. For process concept design, a steady-stage mass balance with an accuracy of 95 % is in general good enough.

Individual atom balances

A second basic element of process concept design is that individual atom inputs are equal to the individual atom outputs. This is because atoms are not destroyed nor made in processes.

This individual atom balance is very useful in checking whether the process concept design is complete; meaning that for all atom types an output stream has been defined. Especially for crude feeds such as biomass containing many mineral atoms besides organic atoms and for crude minerals, such as ores and crude oils, a simple mineral atom check will reveal whether the process concept design is complete, or whether additional output streams are needed for these atoms.

There are three textbook approaches to process concept design. The Douglas approach [1], the Smith approach [2], and the Siirola approach [3]. We describe the main features of these three approaches and take their most useful features to synthesize a

https://doi.org/10.1515/9783111203256-003

process concept design method described in Section 1.2. The selection of the most useful features is made plausible and is based on the author's experience of what is especially useful in process design. He worked 33 years inside Shell, where he witnessed many process designs that made it into commercial scale operation.

Douglas [1] follows a hierarchy step-by-step approach. His first step is to take the decision between batch and continuous operation. As the process elements up front are unknown, Douglas provides a simple heuristic rule based on the process production capacity. Below 5,000 t/y production capacity batch operation is chosen and above 10,000 t/y a continuous operation. The second step is defining the input and output structure. He does this by first drawing the process boundary as a rectangular box to which all input streams and all output streams are to be defined and then draws the input streams and output streams as arrows. The third step is the recycle stream structure, which he combines with a reactor design. Reaction options considered are the laboratory reactor used by the chemist and simple reactor types, plug-flow and back-mixed reactors. The fourth step is the separation design based on phase splits, and the fifth step is the heat-exchanger network design.

Douglas' first step deciding between batch and continuous operations is nowadays not valid anymore. Small capacity processes are now also designed for a continuous operation, while the investment cost is low. This is, for instance, achieved by applying micro reactors and micro separators. Often these designs are more efficient than batch processes. Also, the scale-up to produce more of the same product for product testing as applied in pharmaceuticals is easy. The micro reactor can just be run for a longer time, and in some cases, it can even be applied for commercial scale production.

On the other hand, however, capacity processes far above the 5,000 t/y, such as emulsion polymerization, are often still designed as batch processes, as this allows specifically timed additions of components during the batch time, depending on the reaction progress made.

So, the decision batch or continuous operation should be made much later when essential process steps are defined and the nature of the process design is known.

In this book, we therefore do not treat batch process designs, because of their limited applicability.

The second step, defining input and output and the third Douglas step defining the recycle flow, are attractive first steps to take, as input and output flow and the recycle flow structure require little information and already internal streams sizes and compositions become known, so that they can be used in process design steps to follow. Douglas starts it as a design method with drawing a box that will contain all process steps to be made. Input streams and output streams are connected to this box, so that the box defines the process design boundaries. This box representation with only inputs and outputs is a helpful method in discussing process options with nonchemical engineers, as the focus is first on just getting all necessary inputs and outputs.

Smith [2] also follows a hierarchy of design steps. His onion-layer method starts at the process center where first the reactor is designed, then in the second step separations are defined, the third step is defining the recycle structure, the fourth step is detailed synthesis of the distillation section, and the fifth step is the heat exchange structure.

In Smith's approach, process streams are not known in the early steps and are also not used in his design method. Furthermore, a lot of design effort is needed in each step, while the optimum interactions between unit operations and also the use of streams are not treated in his method.

Siirola's design approach [3] starts with the process input and output definitions, just as in the Douglas method. After that, he describes steps to define functions (which he calls tasks) to connect the inputs to the outputs. The functions are simple transformation tasks, such as reaction, mixing, separation, enthalpy change, and forming (shaping). These transform functions with their inputs and outputs can then be used to connect the process inputs to the process outputs. Each transform function only describes what it does, but not how it is done. Seider introduces Siirola's function approach by using an example [4], so the approach belongs to textbook knowledge.

Siirola does not prescribe a rigid sequence of steps but rather defines functions and streams in an iterative way. He discusses, for instance, the initial requirements of an agent for extraction and later on internal process streams are considered as an extracting agent. In his approach, streams also play an active role. They can be used as a solvent or an extraction agent or a heat carrier.

Siirola's function approach can also be used to get a picture of the process flow sheet, by selecting for each function a conventional unit operation, so that quickly a process concept is made.

In Chapter 2, Siirola' function identification is followed by function integration, resulting in innovative design solutions, not based on established unit operations.

In my career at Shell, I took note of process designs and also how process designs are developed. For several years, I also shared a room with an experienced process engineer, Peter de Blank, who was always willing to explain his way of working. He was a prime example of a reflective practitioner. Here, I share some of my observations on industrial process concept design. I noticed that in industrial process concept design additional solvents are hardly used, but that often-available process streams such as feed streams are used as solvent agents. This has two advantages. No additional storage for the solvent is needed and contamination of streams by the solvent is avoided. I also noted that internal streams are also used as a heat carrier, such as in the FCC process where the fluidized catalyst is used as a heat carrier to transfer heat from the combustor to the cracking reactor.

This industry process concept design experience is taken to select elements of all these three textbook approaches as follows. The outside-in hierarchy approach of Douglas by defining first the input and output streams and then defining recycle streams is taken as the first step. From Siirola, his view on streams as active agents is taken, as well as his iterative approach for defining process functions and using streams. Iterations of the steps are often very useful to find better and more integrated process concepts, when stream sizes and composition are better known. Streams can then be selected as active agents in process concept design. They can, for instance, to be used as a solvent. From Smith, the heat integration method of the whole process is taken.

1.2 Process concept design method

1.2.1 Workflow process concept design overview

The workflow of process concept design presented here is that of Douglas to which Sirola's method of using process streams is added. With the selection of unit operations, the workflow design method finishes. The steps to be taken are numbered. However, there is no need to stick to the sequence of these steps. At the end of taking all steps or even halfway of taking these steps, the resulting concept design can be reviewed, and the design may start with a different sequence of steps.

When the process concept design is finished and it is evaluated using methods of Chapters 5 and 6, the design may be modified to obtain an optimized design.

1.2.2 Process concept design with functions

Step 1: Start with product output stream definition: capacity, composition, and properties

Draw a box. Add output arrows and input arrows for product streams and input streams to this box. The whole process design will be put inside that box by connecting input streams via process steps to output streams. The input streams and output streams may be revisited later on for their size and composition. Also, additional input and output streams may be added.

Industrial designers of a new process start with the required product output (ton/year) of the process and take this as a given.

The production capacity number may stem from a first marketing analysis. If there is not a given, they take a typical value for the plant capacity for that particular company or industry that for bulk chemicals a simple figure, such as 100,000 t/y will do for the time being. When later in the process development of a real number is obtained, the output stream is adjusted to that number, and all other stream sizes are changed by the same adjustment factor.

The product composition also has to be known before the design can start. For existing product specifications, these are available from the marketing department. Bulk chemical companies often publish an industrial chemicals book, such as the Shell Industrial Chemicals Products handbook. It contains product specifications for each product. These specifications are not only on composition, but also on properties such as (the absence of) color with their ASTM test methods. The product composition may be added to a stream composition table.

Step 2: Determine input streams

Then the required process feed inputs are to be determined. As a start, assume that no by-products are formed. The amount of chemicals needed for the product follow then directly from the stoichiometric equations for the reactions.

If no reactions are involved, then the feed stream components can be directly determined from the mass balances over the product and feed streams.

The feed components required may not be available as pure streams. The available feed streams may contain impurities, which have to be separated in the process design either directly or later from the crude product streams. In any way, additional output streams are to be defined containing these impurities.

Step 3: Finalize the output and input stream definitions with mass and atom balances

By a few iterations, all output streams and input streams can be defined. A final check is then made by adding all output streams expressed in mass per year and also adding all input streams, expressed in mass per year. The total mass of the output streams should then match the total mass of the input streams.

An additional check can be made by making atom balances, as atoms are not made or disappear in chemical processes. The amount of an atom type entering the process will also leave the process. So, atom balances are an additional way of checking the process inputs and outputs. In particular, when a crude feed such as biomass is involved, which contain several atoms such as K, Ph, Na, S, and Mg, such a balance often reveals that more output streams are to be defined.

Step 4: Function transformation blocks determination

Seider describes in his textbook the use of functions as a useful step in process concept design. Functions are defined as transformation of input streams to the output streams focused on what the transformation should be (and not how it should be done) [4]. For the first version of the concept, design simple transformation blocks can be used for the list of transformation options: mass movement mixing, reaction, separation, enthalpy change, forming obtained from Siirola [3]. Table 1.1 and Figure 1.1 illustrate the use of functions in process concept design.

Table 1.1: Transformation functions and unit operation options.

Transformation function	Unit operation examples
Mass movement	Gravity, pump, centrifuge
Mixing	Static mixer, mechanical mixer
Reaction	Reactor
Separation	Distillation, extraction, absorption
Enthalpy change	Evaporation, condensing, heat exchanger
Forming	Extruder, Prill tower

In a second version, specific unit operations can be selected for these transformation blocks.

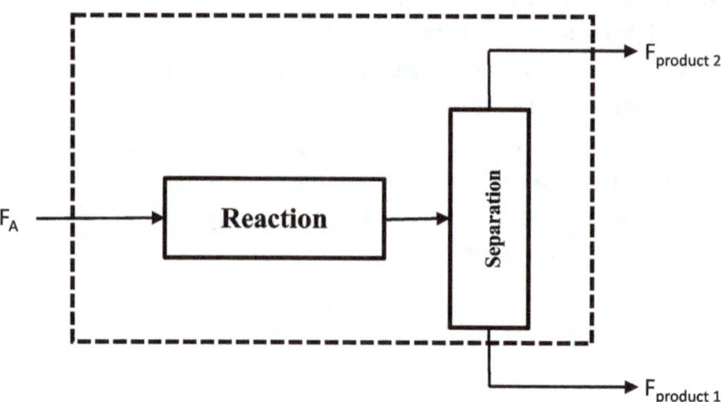

Figure 1.1: Process concept design.

Step 5: Combine reaction engineering and separation design

For reactions involving equilibrium reactions, the conversion of feed to product is limited. For that case. a separation step is needed to recover unconverted feed from the product and recycle the unconverted feed component(s) back to the reactor inlet. In this way. the product yield on the process feed can still be 100 %,

For reactions where the product reacts further to unwanted components. the single pass conversion of the feed should be limited. The product should have a short residence time in the reactor and in separation the product should be quickly removed from the reactive conditions. The unconverted feed components should be recycled back to the reactor.

Step 6: Use of streams as solvents

Streams are not just entities connecting process units but can be seen as entities to be used for certain purposes. This purpose can be a solvent to keep components as a liquid for reactions, or for ease of heat transfer and transportation, or as an absorption or extraction medium. The stream to be used as a solvent can be a feed stream, or product stream, or an internal process stream.

It is highly beneficial for process economics and also for keeping safety measures simple to use an available process stream as a solvent, rather than to introduce an additional solvent to the process. An additional solvent will need additional storage, hence higher investment cost, and additional safety and environmental measures.

Step 7: Unit operation selection

For each functional transformation block, a unit operation is selected, so that a complete process design can be made. Reactor selection is first about reactor residence time distribution and mixing type selection in view of by-product formation reduction. Options are plug-flow, back-mixed, and combinations. If undesired by-products are formed by a consecutive reaction, then plug flow is needed in combination with low single pass conversion. The reactor outlet is treated by a separation step, so that unconverted feed components are recycled back to the reactor.

This is then followed by reactor type selection. Options are, depending on solid phase presence, fixed bed, fluid bed, and slurry. The reader is referred to chemical reaction engineering textbooks for selecting a reactor type. For selecting separation unit operations, various selection guidelines, also called heuristics, are available. These are treated in Section 1.2.3.

A more detailed set of heuristics is provided by Douglas [1]. He starts with the phase of the reactor outlet. If that is a liquid, then a liquid separation such as distillation should be considered to obtain the product as a vapor stream. If the reactor outlet has two phases, then first a phase splitter is chosen. The vapor may be cooled to form a liquid. Available cooling water should be used as the coolant.

The reason for this heuristic about selecting the product stream as a vapor stream from distillation is that by-products are often heavy boilers. By taking the product as vapor stream from distillation removes these heavy boilers from the product stream.

If the liquid contains mainly reactants, then that stream is recycled back to the reactor. The product in the remaining vapor stream can then be recovered from the process, often after a purification step. If the reactor outlet is a vapor, then cool it down to obtain a two-phase stream. Then the previous heuristics are used to recover a product stream and recycle unconverted reactants back to the reactor. The background to these heuristics is that distillation and vapor-liquid separation are a low investment cost with low energy demands and should be considered before other separation options.

If simple distillation is not feasible due to azeotrope formation, then dual columns operating at different pressures is the preferred option over using an additional solvent. An additional solvent means additional storage cost and a risk of contamination of the product stream.

If distillation is not feasible and the contaminant is a small fraction, i. e., less than 10 % then adsorption can be considered as a separation step. Chapter 19 provides a concept design method for adsorption.

If a very product high purity is required, then crystallization can be an option to consider.

Step 8: Design unit operations and streams

All unit operations are now designed for conditions such as temperature, process, and stream compositions for input, output, and internal unit operations.

For reaction systems containing consecutive reactions, or equilibrium reactions, and for distillation involving azeotropes or close boiling systems, reactor design and distillation design should be combined with recycle streams to get the optimum solution. In most cases, this is done with a computer software flow sheeter package.

If the feed stream contains large amounts of impurities, such as an air feed stream with 80 % nitrogen, then separating the impurities first is economically attractive. Then all subsequent process steps do not have treat this high flow of impurities so that they become much smaller. If the impurities act as a catalyst poison or cause separation problems with the product, then also the impurities should be removed up front of the other process steps [1].

Step 9: Size energy services

All required energy supply and withdrawal (cooling) for conditions and amounts are determined for each unit operation. Heat exchangers are designed using a log temperature difference, and simple heat transfer correlations to determine the Heat Transfer Area (HTA), which for cost estimates are detailed enough.

For process concept design, it is assumed that a central utility supply is available to provide utility services such as steam, cooling water, and electricity.

Energy integration of unit operations to save energy is in general not carried out during concept design, but in the feasibility or development project stages.

Step 10: Review process design

Steps 1–9 suggest that process concept design is following simply sequentially these steps. In reality, several iterations will be made over a few or even over all steps. When the concept design is judged to be good enough, it will be challenged by others and smaller or bigger changes will be made.

Table 1.2 summarizes steps 1–10.

Table 1.2: Process design steps.

Process design step	Attributes
Step 1: Start with product output stream definitions	Capacity, compositions, specifications Client destinations
Step 2: Determine input streams	Capacity compositions, specifications Provider sources
Step 3: Finalize output and input stream definitions	Check mass balance and atom balance
Step 4: Function transformation blocks determination	
Step 5: Combine reaction engineering and separation	Aim for minimal by-product formation
Step 6: Use of streams as solvents	Avoid additional solvents
Step 7: Select unit operations	Select best types
Step 8: Design unit operations and streams	Size with simple correlations
Step 9: Size energy services	Size with simple correlations
Step 10: Review process design	

1.2.3 Unit operation separation selection

1.2.3.1 Basics selecting molecular separation methods

Molecular separation is about separating different molecules from each other. Their molecular differences also mean that they have different physical properties that can be exploited for separation. Selecting a separation method is in most cases combined with process concept design. That subject is treated in Chapter 2.

Here, just a quick selection method is provided using a simple hierarchy based on plausible reasons for process design based on economics and backed up by the textbooks of Douglas on process concept design [1] and of de Haan on separations [5]. The latter provides a list of separation methods applied at commercial scale, which is used in deriving the selection method of this section.

Rationale for the separation hierarchy

The overall rationale for the hierarchy selection method of this section is that additional components for separation should be avoided or at least be minimized. The reason is that the use of an additional component means additional storage vessels, additional process recovery steps constructions, and additional safety, health, and environmental measures. So, additional supply components mean higher cost. Separations that need no additional components are therefore first to be considered. Separations that need additional process steps are considered thereafter. Based on this reasoning, Table 1.1 is made with a hierarchy for selection 1–6.

Table 1.3 provides a list of molecular separation methods that have been applied at commercial scale [2]. The methods are listed from top to bottom as a selection hierarchy.

Table 1.3: Selection hierarchy for molecular separation methods.

Separation method	Additional component	Additional process steps
1 Distillation	–	–
2 Absorption	Absorbent	Absorbent treatment
3 Crystallization	Washing component	Filtration + washing
4 Liquid-liquid extraction	Solvent	Solvent recovery
5 Adsorption	Adsorbent	Adsorbent recovery
6 Membrane		Membrane cleaning

1.2.3.2 Distillation first

In molecular separation, other criteria are also to be considered such as required purity to be achieved, and yield of desired product on the feed stream. So, applying the hierarchy of the steps in Table 1.1 is just an aid in the whole separation selection method. The first step to consider is distillation.

The rationale for this hierarchy is that distillation does not require an additional separation component. It also does not require an additional process step.

Distillation can be a single stage flash or a multistage column. It can be operated at subatmospheric pressure to enhance its separation effectivity, or at high pressure, for instance, to facilitate energy savings by heat integration with other process sections.

The distillation can be executed in a single column or if more components have to be separated then a train of distillation column can perform this separation. For such a train, dividing wall column distillation is an additional option to save energy and investment cost.

If straight distillation cannot perform the required separation, for instance, because an azeotrope is formed, the solvent assisted distillation can be considered. However, then it has the same disadvantage as the other separations, so then the other separation methods should also be considered, and an elaborate evaluation will be needed to select the best separation option.

1.2.3.3 Consider absorption second

Absorption is similar to distillation. It needs, however, a liquid absorbent with an affinity for the component to be separated from the other component. The absorption requires a column with a packing similar to distillation.

The absorbent loaded with the separated component has to be treated in a downstream process.

So, this process step needs an additional component, which needs a supply system.

1.2.3.4 Consider third crystallization

If components cannot be separated by distillation or absorption, for instance, because their vapor-liquid equilibria are very close to each other and absorption is not feasible either, then crystallization is to be considered. This option is often applied if isomers have to be separated from each other.

By cooling down the stream with the mixture to the crystallization temperature, crystals are formed. In a second step, the crystals are separated from the liquor by settling or filtration. To obtain a high purity product, any liquor with the other component should be removed from the crystals. This can be done by a washing step. So, then as an additional component, a washing liquid is involved.

1.2.3.5 Consider fourth: Liquid-liquid extraction

In liquid-liquid extraction, a selective extracting agent, called a solvent, is to be found, which selective solvent extracts the component from the mother liquor to the solvent. In a second step, the solvent is separated from the component, often by distillation and then recycled back to the extractor.

Liquid-liquid extraction and crystallization are similar in their additional components and process steps. So, often both separations are evaluated next to each other in some detail.

1.2.3.6 Consider fifth adsorption

In adsorption, one component is selectively adsorbed onto a solid surface. The solid surface is in most cases a porous material with a high internal surface area.

The process equipment is in most cases a packed bed. When the porous material is loaded, the feed is stopped and the component is recovered from the adsorbent either by stripping with a stripping agent, or by heating the porous material.

In most cases, the feed is a gas, but adsorption can also be applied for liquid feeds.

In many designs, two or more fixed adsorption beds are located in parallel. When one bed is loaded, the feed is switched to the second bed. The first bed can then be stripped of the adsorbed component.

Adsorption is used at large scale in purifying hydrogen from carbon dioxide and water, as a downstream process step after the steam reforming of methane reaction step. It is also used to remove trace amounts of a certain component from a stream.

1.2.3.7 Consider sixth: membrane separation

Membrane separation has developed enormously in the last 60 years for all kinds of separation applications. It produces a high purity product stream. The reason that it is the last method in this hierarchy is that for most applications it has three disadvantages:

a) The yield of product on the feed stream is limited;
b) The investment cost for large capacity applications is high;
c) It requires a membrane cleaning step.

Ad a: A high yield of product from the feed stream is often impossible because of a resulting high osmotic pressure drop over the membrane.

Ad b: The investment cost for large capacity applications is large because the investment cost scale is nearly linear with capacity.

Ad c: In many applications, the membrane fouls and a cleaning step is needed. This introduces an additional cleaning step with an additional resulting stream to be treated.

1.2.3.8 Limitation to the selection method

The limitation to the presented selection method is that it is based on a plausible hierarchy, but it is not based on scientific research.

1.3 Case example: Process concept design bisphenol A

Bisphenol A is a bulk chemical produced for over 50 years. It is an industrial intermediate used for epoxy resin manufacturing and for manufacturing. Several processes are described in literature [6–10].

The required bisphenol A product output stream is set at a typical value of 100,000 t/y.

The required input streams follow from the stoichiometric reaction equation:

$$\text{Reaction:} \quad 2\,\text{Phenol} + 1\,\text{Acetone} = 1\,\text{Bisphenol Acetone} + 1\,\text{water}$$

$$2\,C_6H_6O_1 + 1\,C_3H_6O_1 = 1\,C_{12}H_{16}O_2 + 1\,H_2O$$

Molecular masses are:

$$M_{ph} = 94; \quad M_{ac} = 58; \quad M = 228; \quad M_{H_2O} = 18$$

The input streams phenol and acetone can now be determined. Also, the additional output stream water can be quantified. The resulting output and input streams are:

$$\text{Output product Bisphenol A stream,}_{ph} = 100{,}000\ \text{t/y}$$

$$\text{Output stream water} \qquad F_{water} = 7{,}900\ \text{t/y}$$

$$\text{Total output} \qquad F_{outtot} = 107{,}900\ \text{t/y}$$

$$\text{Input stream phenol} \qquad F_{phenol} = 82{,}500\ \text{t/y}$$

$$\text{Input stream acetone} \qquad F_{acetone} = 25{,}400\ \text{t/y}$$

$$\text{Total input} \qquad F_{intot} = 107{,}900\ \text{t/y}$$

The reaction function is now needed as well as at least one separation function to separate the product from water. The simplest process design is now given by Figure 1.2 with a reaction function and a separation function and 5 streams; water, phenol, water-bisphenol A mixture, water, and bisphenol A.

The melting point of product bisphenol A is 158 °C, so a solvent is needed to be able to treat it as a liquid in the reactor and in the separator. The available streams as a potential solvent are phenol, acetone, and water. Bisphenol A is miscible with phenol over a wide range. So, phenol is the first option of choice. If phenol is in the reactor at a high ratio to acetone, then deep conversion of acetone is feasible, by which acetone does not need to be separated from water and the product. So, phenol is chosen as the solvent for the product.

The melting point of phenol is 40.5 °C. It is provided as a liquid from a heated storage. The atmospheric pressure boiling point of phenol is 182 °C. The boiling point of bisphenol A at 0.02 bar is 252 °C; phenol can be easily distilled from bisphenol A.

The resulting functional block flow process scheme is shown in Figure 1.3.

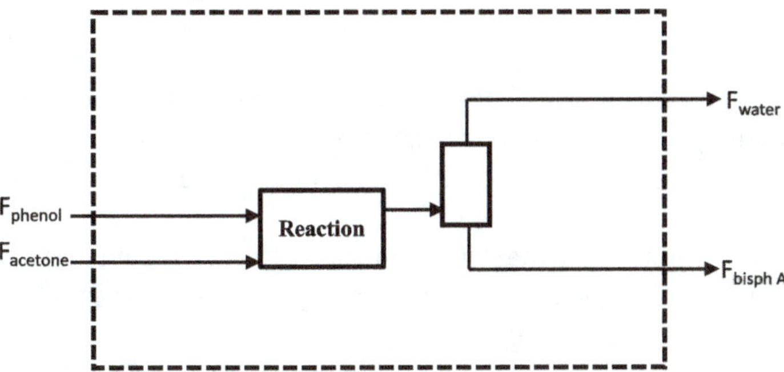

Figure 1.2: Bisphenol A process concept design no 1.

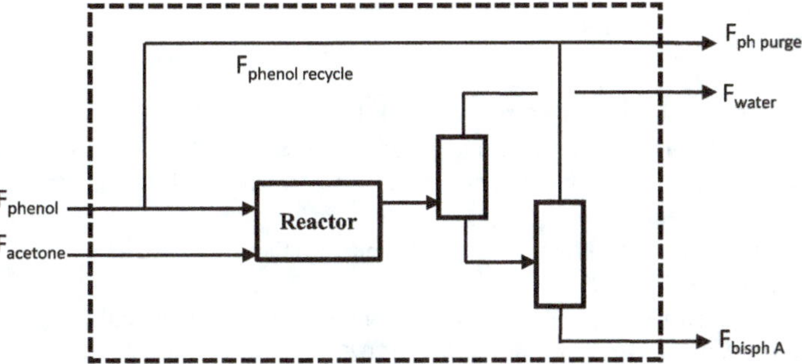

Figure 1.3: Bisphenol A process concept design with phenol as solvent.

For the block reaction, several unit operations can be chosen. One option is an ion exchange resin to act as an acid catalyst [9]. The reaction unit operation is then a fixed bed reactor.

For the separation blocks, distillations are the obvious choices.

The product is sold as a granular solid. So, the liquid product can be solidified by cooling it and forming solids, for instance, in a spray cooling tower. So, this will be added later in the project development stage or be added in the concept stage.

It should be noted that in early concept design pumps are not included and also the energy supply of the unit operations is not defined yet.

Furthermore, it is assumed that the wastewater output stream destination wastewater treatment plant is acceptable, as this wastewater will contain traces of phenol that is not a realistic assumption. An extraction section will be needed to remove phenol traces.

In the phenol recycle stream, trace components from the feed and from by-product reaction may build up, so a phenol purge stream is envisaged to limit the trance components build up.

In the development stage, recovery of pure phenol from the purge stream will be investigated to reduce the cost of phenol intake and reduce the waste stream.

1.4 Case example: BPA ortho-para separation selection

In the commercial scale production of Bis-Phenol-Acetone (BPA) in the reaction section, two isomers are formed para,para (p,p) and the ortho,para (o,p) isomers. The p,p isomer needs to be produced at a very high purity. This high purity can only be achieved by crystallization and not by other molecular separations.

The p,p is selectively crystallized from the solution in phenol (phenol is both the feed component as well as the solvent in the BPA process). The crystals are then separated from the o,p BPA, and phenol by filtration [10].

So here, no additional solvent is needed. Only two additional steps are needed to recover phenol from the p,p BPA and from the o,p BPA.

Bibliography

[1]   Douglas JM. Conceptual Design of Chemical Processes. New York: McGraw-Hill; 1988.
[2]   Smith R. Chemical Process: Design and Integration. Hoboken: John Wiley & Sons; 2005.
[3]   Siirola JJ. Industrial applications of chemical process synthesis. In: Advances in Chemical Engineering. Vol. 23. Academic Press; 1996. pp. 1–62.
[4]   Seider WD, Seader JD, Lewin DR. Product & Process Design Principles: Synthesis, Analysis and Evaluation. Hoboken: John Wiley & Sons; 2009.
[5]   De Haan AB, Eral H, Schuur B. Industrial Separation Processes: Thermal Unit Operations and Mechanical Unit Operations. Walter de Gruyter GmbH & Co KG; 2025.
[6]   Mitsubishi chemical company, Introduction to MCC BPA process, Resourced 5—2023. https://www.m-chemical.co.jp/en/petrochem-license/technologies/pdf/Introduction_MCC_BPA_Process.pdf.
[7]   General Electric, Process for making bisphenol A, patent EP 0210366 B, 1990.
[8]   Altuwair I. Production of bisphenol A (BPA) by green technology. Engineering Technology Open Access Journal. 2018;1:72.
[9]   Neagu L. Synthesis of bisphenol A with heterogeneous catalysts. Sourced: 4 Oct 2024. https://www.collectionscanada.gc.ca/obj/s4/f2/dsk2/tape17/PQDD_0005/MQ37971.pdf.
[10] Kiedik M, Chruściel A, Hreczuch W. The MEXEO bisphenol A (BPA) technology (case study). In: Industrial Arene Chemistry: Markets, Technologies, Sustainable Processes and Cases Studies of Aromatic Commodities. Vol. 3. 2023 Jun 6. pp. 1359–84.

2 Process intensified design

2.1 Basics process intensification

2.1.1 Process intensification definitions

Keil discusses in his magnificent review article various PI definitions and concludes with the definition of the European Roadmap of Process Intensification: "Process Intensification is a set of often radically innovative principles ("paradigm shift") in process and equipment design, which can bring significant (more than factor 2) benefits in terms of process and chain efficiency, capital and operating expenses, quality, wastes, and process safety" [1]. The same definition appears in the textbook on process intensification by Stankiewicz [2].

The remarkable thing about this definition (as with all other PI definitions reported by Keil) is that it does not specify the reference from which the benefits are made, or on what existing paradigm the shift is made. The most obvious existing paradigm is of course that of process designs based on conventional unit operations.

With this insight, Harmsen [3] refined the definition of process intensification as:

Process Intensification is a set of radically innovative process design principles, which can bring significant benefits in terms of efficiency, cost, product quality, safety, and health [1, 2], over process designs based on conventional unit operations [3].

There are three different approaches that are described to obtain process intensified designs [4]:
– The research approach
– The intensified unit operations selection approach
– The function integrated process design approach

2.1.2 Research approach to generate intensified unit operations

The research approach to process intensification is described in detail by Stankiewicz in his textbook [2]. This approach uses four principles towards process intensification, which then can be used to initiate research into new unit operations. The four principles are:
– Perfect reactions
– Ideal experiences for molecules
– Optimized driving forces, resistances, and interfaces
– Synergies

With perfect reactions, Stankiewicz means that the reacting molecule has the perfect kinetic energy for reaction. So, Stankiewicz moves here away from a macroscopic treat-

https://doi.org/10.1515/9783111203256-004

ment of molecules with the Bolzman energy distribution around the mean and treats individual molecules.

With ideal experiences of molecules Stankiewicz means that each molecule has the ideal experience. This means, for instance, in his view that all molecules have the same ideal residence time; hence, plug flow, but then on the molecular scale and not on the macroscopic scale.

With optimized driving forces, resistances, and interfaces he means to optimize the combination of driving forces, resistances, and interfaces. This means increasing the interfaces (surface-to-volume ratios), for instance, by microchannels and micro droplets. The resistances then drop anyway.

With synergies, he means synergistic effects of bringing multiple functions together in one component. As an example, he mentions the monolithic stirrer reactor, a result of research at the Delft University of Technology. The reactor concept combines the heterogeneous catalyst function with the stirrer.

2.1.3 Established intensified unit operations categories

The intensified unit operations are categorized by Stankiewicz in his textbook in four process intensification domains [2]:
– Spatial
– Thermodynamic
– Synergy
– Temporal

For each domain, many novel unit operations are described in detail in his textbook. Some of these unit operations are still in academic research. These are outside the scope of this design book. However, others have been applied at the commercial scale and can be used to select in process design.

2.1.4 Function integrated design approach

The function integrated design approach stems from the process synthesis method. It started with a book: Process Synthesis by Rudd, Powers, and Siirola in 1973 [5]. Process synthesis is carried out in five steps. In each step, differences between the process input and the process desired output are eliminated by a function. Table 2.1 shows most used functions.

This method became known when Siirola published results of this method in industrial process designs and, in particular, when Eastman Chemical published their commercial scale methyl acetate process in which 11 conventional unit operations were re-

Table 2.1: Process synthesis steps.

Process difference elimination step	Function
1 molecular	Chemical reaction
2 distribute chemicals	Mixing
3 composition	Separation
4 temperature, pressure, phase	Change (by transfer)
5 integrate tasks	

placed by an extractive reactive distillation column, which required a 20 % investment cost and energy compared to the conventional design [6, 7].

The Siirola's approach was successful inside Eastman Chemicals. The methyl acetate design was implemented at a commercial scale with a capacity of 200,000 t/y of methyl acetate [6, 7]; however, still, his complete method is not taken up by academic teachers in their curriculum so far.

A major item of his method however has been taken up and that is the use of functions (or tasks, as Siirola calls them) rather than directly taking existing unit operations. What is also taken up is the use of multifunctional solutions such as reactive distillation, reactive extraction, mixing in pumps, etc. Seider, for instance, describes in his textbook the function design method and applies it to a vinyl chloride process design [8]. Babi refers to the method [9] and also Demirel provides a process concept design method using functions. He calls the functions basic building blocks [10]. Linke also uses in his first steps of process concept design functions, which he calls tasks [11], like Siirola. Also, Pistikopoulos [12], Freund [13], and Sargent [14] refer to the use of functions in process concept design.

2.1.5 Example: Fluid Catalytic Cracking (FCC) process

In hindsight, it is likely that the Fluid Catalytic Cracking (FCC) process designed in 1943 to produce large amounts of petrol for fueling second world war transport, was designed first in functions for catalytic cracking, and catalyst regeneration, followed by deciding to perform both functions by fluid bed reactors, which facilitated a continuous process operation with very high capacity at low investment cost and no external energy requirements.

The catalytic cracker reactor was designed as a riser fluid bed reactor in which catalyst and oil vapor were transported upwards at high velocity (velocity of gas and catalyst of 10–20 m/s, while cracking occurred. By cyclones, the cracked vapor was separated from the spent catalyst.

The spent catalyst on which coke was deposited was sent to the second fluid bed reactor fed with air. The coke oxidized to carbon dioxide and the regenerated catalyst was transported by a standpipe under dense fluid bed conditions back to the riser reactor.

The heat generated by the coke combustion in the generator reactor was transported to the endothermic cracker riser reactor, avoiding heat exchangers between these reactors.

The idea of using fluidization came from MIT. A pilot plant was built and operated in 1940. Then a commercial process was designed and came into operation in 1942. The commercial scale process was an immediate success [15, 16].

It is worthwhile to mention that at that time no computers were available. No computational fluid flow modeling was available. Hand and slide rulers performed all calculations.

After the war, all oil majors applied the process, and it is still in operation in my refineries. Modeling of the reactors also started at many universities and is still studied.

This FCC process history reveals some major points for chemical process design and development in general.

A)   A concept design first and then it is followed by Research & Development
B)   The concept design with functions is followed up by choosing reactor types for each function.
C)   The reactors are integrated with each other by fluidized mass flows, which also transport heat.

2.2 Process intensification concept design

2.2.1 Process intensification for inherent safer design

The concept of inherent safer design was introduced by Trevor Kletz after the Flixborough disaster, and his expression sums up the basic principle: *"What you don't have cannot leak or burn"* [17]. There are four strategies for achieving inherent safer designs [18]:

1.   Minimize: To reduce the quantity of material, energy, or energy density contained in a manufacturing process or plant
2.   Substitute: The replacement of a hazardous material or process with an alternative that reduces or eliminates the hazard
3.   Moderate: Using materials under less hazardous or energetic conditions, or designing the plant to mitigate the impact of an incident arising from the hazard
4.   Simplify: Designing and/or operating the process to reduce or eliminate unnecessary complexity in order to reduce or eliminate the chemical hazard.

Since no process can be considered inherently safe, the accurate term used here is "inherently safer design." Hence, it is always relative to another reference process.

Process design by process intensification reduces by-product waste and energy requirements, hence it follows strategy number 1. It also simplifies the process design by having less process steps and process equipment; hence, it follows strategy step number 4.

2.2.2 Step-by-step process intensification design

Step 1: Define inputs and output streams as described in the previous chapter.

Step 2: Define functions to connect input streams to output streams. The functions are represented by blocks. The functions are connected by streams represented by arrows.

Step 3: Use available streams of the established block flow diagram for specific actions such as stripping and extraction. These stream actions are also represented in the block flow diagram.

Step 4: Combine as many functions as possible in a single vessel. Table 2.2 shows which functions can be combined. Section 2.3 shows an industrial implemented example of combining functions.

Step 5: Select for the remaining functions established unit operations. Established process intensified unit operations, such as those provided by Stankiewicz, should be investigated first.

Table 2.2: Function combination options.

Function	Mass movement	Mixing	Reaction	Separation	Enthalpy change	Forming
Mass movement	x	X	X		x	X
Mixing	x	X	X		x	X
Reaction	x	X	X	X	x	X
Separation	x		X	X	x	X
Enthalpy change	x	X	X	X	x	X
Forming	x	X	X	X	x	X

Flow sheet modeling will facilitate this step-by-step concept design. It will show the feasibility of function combinations. The flow sheet model will also facilitate the process optimization.

2.3 Case example: Biomass liquefaction process function integration

The company BTG-BTL developed a biomass liquefaction process, which was successfully implemented at commercial scale and published by Harmsen [3]. Figure 2.1 shows a block-flow diagram of the process highlighting the process function integration items. Here follows a description of the functions and their integration.

In the fluid bed rotating cone reactor, biomass is rapidly mixed with hot sand. This causes the biomass to rapidly heat up and to gasify. The resulting gas acts like a fluidization gas and no additional fluidization gas is added to the fluid bed. The gas and sand move over the rotating cone upwards and leave the reactor.

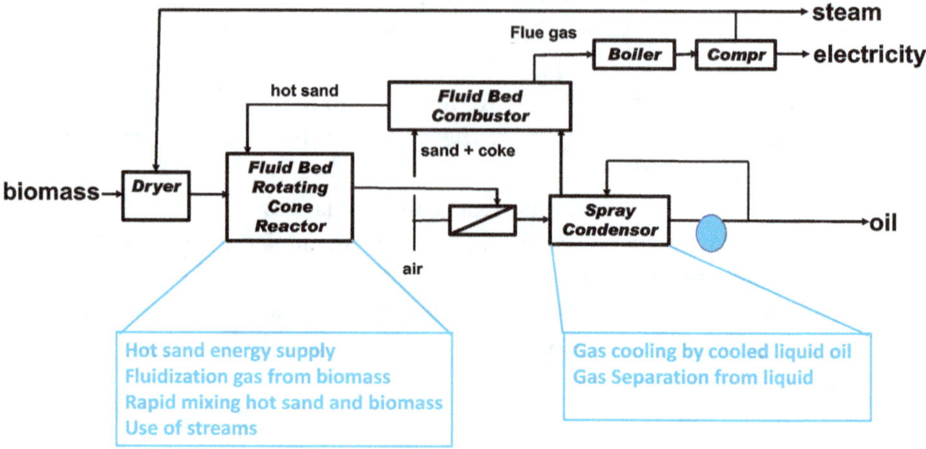

Figure 2.1: Process function integration design biofuel process BTG-BTL.

Bibliography

[1] Keil FJ. Process intensification. Reviews in Chemical Engineering. 2018 Feb 23;34(2):135–200.

[2] Stankiewicz A, Van Gerven T, Stefanidis G. The Fundamentals of Process Intensification. John Wiley & Sons; 2019 Sep 16.

[3] Harmsen J, Process VM. Intensification: Breakthrough in Design, Industrial Innovation Practices, and Education. Walter de Gruyter GmbH & Co KG; 2020 Jul 20.

[4] Harmsen J, Verkerk M. A new approach to industrial innovation. Chemical Engineering Progress. 2021 Mar 1;117(3):50–3.

[5] Rudd DF, Powers GJ, Siirola JJ. Process Synthesis. Englewood Cliffs, NJ: Prentice-Hall; 1973.

[6] Siirola JJ. Industrial applications of chemical process synthesis. In: Advances in Chemical Engineering. Vol. 23. Academic Press; 1996 Jan 1. pp. 1–62.

[7] Siirola JJ. An industrial perspective on process synthesis. In: Inarched Symposium Series 1995. Vol. 91. No. 304. New York, NY: American Institute of Chemical Engineers; 1971–2002. pp. 222–34.

[8] Seider WD, Seader JD, Lewin DR. Product & Process Design Principles: Synthesis, Analysis and Evaluation (With CD). John Wiley & Sons; 2009.

[9] Babi DK, Holtbruegge J, Lutze P, Gorak A, Woodley JM, Gani R. Sustainable process synthesis–intensification. Computers & Chemical Engineering. 2015;81:218–44.

[10] Demirel SE, Li J, Hasan MF. A general framework for process synthesis, integration, and intensification. Industrial & Engineering Chemistry Research. 2019;58(15):5950–67.

[11] Linke P, Kokossis A, Van den Berg H. Process synthesis/integration. In: Re-engineering the Chemical Processing Plant–Process Intensification. New York: CRC Press/Marcel Dekker, Inc.; 2003. pp. 409–46.

[12] Pistikopoulos EN, Barbosa-Povoa A, Lee JH, Misener R, Mitsos A, Reklaitis GV, Venkatasubramanian V, You F, Gani R. Process systems engineering–the generation next? Computers & Chemical Engineering. 2021;147:107252.

[13] Freund H, Sundmacher K. Towards a methodology for the systematic analysis and design of efficient chemical processes: Part 1. From unit operations to elementary process functions. Chemical Engineering and Processing: Process Intensification. 2008;47(12):2051–60.

[14] Sargent RWH. A functional approach to process synthesis and its application to distillation systems. Computers & Chemical Engineering. 1998;22(1–2):31–45.

[15] Squires AM. The story of fluid catalytic cracking: The first "circulating fluid bed." In: Circulating Fluidized Bed Technology Pergamon; 1986 Jan 1. pp. 1–19.

[16] Palucka T. The wizard of octane: Eugene Houdry. Invention & Technology. 2005;20(3). Archived from the original on 2008-06-02. Retrieved 2008-05-10.

[17] Kletz TA. Inherently safer design – The growth of an idea. Process Safety Progress. 1996;15:5–8.

[18] CCPS. Inherently Safer Chemical Processes: A Life Cycle Approach. 3rd ed. 2020.

3 Modular process design and skid mounted prefabrication

3.1 Basics modular design and skid mounted prefabrication

There are a large number of publications, which use the term modular process design, prefabrication, skid-mounted and distributed design in all kinds of ways and in different contexts. The text below should clarify this field.

Location in engineering, procurement and construction innovation stage

Modular process design and prefabrication is all carried out in the Engineering, Procurement and Construction innovation stage (EPC). So, it is carried out after the concept stage and after the development stage of an innovation project. Any concept design, be it conventional or process intensified, can be designed as a modular design in this EPC stage—prefabrication, meaning pre-construction at the EPC company site, instead of direct construction at the site where the process is to be in operation. The prefabricated process is transported in modular pieces to the manufacturing site and there the modular pieces are re-connected.

Detailed engineering

Detailed engineering means that every process item is designed in size and construction material, in which all process control and instrumentation are defined and that the structure supporting all process equipment and piping is designed. The number of elements to be defined in detail engineering design are in the order of 10,000. It is an enormous effort. For instance, defining all piping, which connect all major equipment with each other, takes 20 % of the total investment cost.

Procurement

Procurement means procuring all major process equipment.

Construction

Construction means the actual fabrication of the process in which all major equipment is installed and connected with pipes and control systems.

Often detailed Engineering, Procurement and Construction (EPC) are all carried out by an EPC contract company in close cooperation with the manufacturing company.

Modular design definition

Modular design means that the process design is split up in modules. Each module is designed and then the resulting design modules are standardized. This means that the

https://doi.org/10.1515/9783111203256-005

design can be used again and again for constructing the same process, without any design effort. It also means that if one module of the design is improved, that the other modules do not need a change.

Prefabrication definition

Prefabrication means that the process is first constructed at the Engineering Procurement Construction (EPC) site. Also, dry testing to validate the technical integrity and proper function of the instrumentation and control is carried out at the construction site. After that, it is deconstructed in pieces, transported to the manufacturing site, and reassembled. Ater that, the process starts up to produce the desired product. So, prefabrication differs from construction at the manufacturing site. The latter is called stick built.

Skid-mounted and container mounted construction definition

To facilitate transportation, the whole process is constructed inside a container or the process is skid mounted. This means that a support structure is attached to the process. If modular design is the base for the prefabrication, then often each module is in its own skid mounted, facilitating transport and there is easy reassembly of the module at the manufacturing site.

Distributed modular design combined with prefabrication

When several small capacity processes for the same purpose are installed at different locations by the same technology owner, then that is often called "distributed processes or distributed plants." Each process plant may be installed sequentially with some time in between each installation. If that difference in time is long enough, say longer than a year, then learning points from a plant in operation can be used to optimize some modules of the process design. In this way, the learning curve can be quickly run down, reducing the investment and operating cost of each subsequent process implemented [1–3].

3.2 Modular design and prefabrication design methods

3.2.1 Introduction

Modular process design and also prefabrication can be applied to any process. However, it is particularly attractive if many of the same processes are needed.

Small capacity process designs are applied when large capacity processes would mean very large transportation cost, such as when the feed is from small capacity sources, such as natural gas from urban waste pits, or biomass waste, or to avoid transport of dangerous chemicals, by producing the dangerous component at the location where it is needed (such as chlorine), or when a small production capacity is needed anyway, such as for pharmaceuticals. Bieringer provides an overview of implemented modular designs in chemical companies [4].

3.2.2 Combining process intensification with modular design

Process intensification methods applied in the concept design often benefit the economics of small capacity processes in at least two ways. The first way is that process intensification based on function integration reduces the investment cost of processes over conventional process designs, see Chapter 2. The second way is that for process intensified unit operations such as high heat exchange reactors, and reverse flow pipe reactors the investment cost scale with a higher exponent than the 0.6 value. Which means that the penalty for small capacity process is less severe.

3.2.3 Economics of small capacity modular design processes

The investment cost per amount of product for small capacity processes, however, will become very high if conventional process design and construction methods are applied. The total investment cost per ton of product is related to the capacity with the power −0.4 (this follows directly from the power 0.6 law relating the investment cost with the production capacity (ton/a) to the power 0.6). The reason of this 0.6 cost curve is three-fold. The first is that construction material needed increases less than the power one with increasing size. The second component is that several items of a process are hardly dependent on the production capacity, such as piping and concrete floors and cable connections. The third is that detailed engineering design costs are hardly dependent on the production capacity, as the detailed design effort is the same, whether the item is small or large. There is however a new design and construction method that breaks the 0.6 rule and that is the modular design in combination with skid mounted design.

When a conscious decision is taken to have several small scale processes distributed over various locations rather than one central large process, then this is called distributed design.

When many of the same modular small-scale processes are needed and the processes are implemented with sufficient time in between to obtain learning points, then the design can be improved. Cost learning curve theory has been applied by Weber [3] to theoretical case studies using cost learning curves from other processes. These theoretical case studies indicate that this additional method or rapid learning can also help to avoid higher cost for the low-capacity small plants.

Patience applied the rapid learning cost curve to a process intensified micro-refinery converting methane (from waste) to liquid fuel using a Fisher–Tropsch process concept and indicated that 100 small processes of 1 bbl/day would then have the same investment cost as a conventional Fisher–Tropsch process of 100 bbl/day [2].

Othman shows that for chemical processes of new products with a rapidly growing market but starting a low capacity needed for early market development, a small capacity modular design and repeated implementation when the market grows can be economically attractive compared to designing a large capacity plant, which is underused in the first years [5].

O'Connor showed in a theoretical desk study that specialty chemicals can be produced with a continuous process intensified modular design with lower cost than a conventional batch process design [6]. This lower cost is most likely mainly due to the change from batch process design to continuous process design, then using a modular design.

From these desk studies, it is already clear that there are opportunities for applying small capacity skid mounted processes, which are economically viable. The economic opportunities grow when the large capacity process involves expensive transportation cost. When the process concept design is a process intensified concept, the investment cost can be further reduced.

Sievers provides a general methodology for evaluating when to apply several (modular) small capacity process designs and when to apply a single large capacity process design [7]. Steyn and van Heerden provide a long list of criteria when to apply modular design with skid mounted construction [8]. These are summarized in Tables 3.1, 3.2, and 3.3. These criteria are not the result of scientific research and Steyn and van Heerden are employees of Owner Team Consultation, so these criteria have limited reliability.

Table 3.1: Economic criteria when to apply modular design and prefabrication.

Repeatability of the design for subsequent plants
Several same units exact duplicates
Stick-built has excessive high cost for labor, housing, and infrastructure work
Schedule time reduction by highly trained and experienced prefabrication technicians
Schedule time reduction by parallel prefabrication and site civil work
Start-up time reduction by precheck at EPC
Preexisting operations at site are not interrupted by stick-built activities

Table 3.2: Specific criteria when to apply modular design with prefabrication over stick-built construction.

Severe weather conditions causing time delays for stick-built
Limited site plot space for stick-built construction
High labor cost for stick-built
Shortage of skills at site for stick-built
Extensive acceptance testing at site
If site permits for construction take a long time to obtain

Table 3.3: Specific criteria when not to apply modular design with prefabrication.

Transport conditions prohibit pre-fabrication
No suitable fabrication houses

They can however be used for evaluating the modular skid-built option for projects in the EPC stage, as the criteria are plausible.

3.2.4 Modular design and prefabrication execution

3.2.4.1 Modular design

Modular design is executed in the detailed engineering step of the EPC stage for a specific process. It may start with an evaluation using Tables 3.1, 3.2, and 3.3 of Section 3.2.3. The next step is to define the production capacity, the product specifications, and the feed specifications. The third step is to split the process concept design into modules, which makes sense from a three-dimensional design point of view, transportation limitations, and from the point of view of minimum connection points between the modules. Details of important other steps are, for instance, provided by LePree [9] and Halfort [10].

3.2.4.2 Prefabrication

Prefabrication of processes is the opposite of stick-built fabrication. Both fabrications are carried out by specialized engineering firms that do not disclose their practices in open literature. The prefabrication is in general executed by mechanical engineers and not chemical engineers. So, the subject is outside the chemical engineering field.

In most cases, prefabrication is combined with modular design. In some publications, modular design is even stated as opposite stick-built fabrication. So, the authors take modular design to be the same as prefabrication.

3.3 Case example: Biofuels modular process design and skid-mounted prefabrication

The company BTG-BTL has developed a process for the conversion of biomass to liquid biofuel. The process concept design is described in Chapter 2. It is a process intensified concept design, combining functions into single pieces of equipment. The capacity of this process is 20,000 ton/y of liquid biofuels.

The EPC company ZETON made for this process a modular skid mounted detailed engineering design, consisting of 9 modules [11].

ZETON also prefabricated the process at their ZETON site in the Netherlands where also the technical integrity and controls were dry tested. After disassembly, the modules were transported by trucks to the production locations.

So far, this prefabrication and implementation of this biofuels have been repeated at least three times. The implementation locations and years were: The Netherlands 2015, Finland 2020, and Sweden 2021 [12].

This case combines process intensification concept design, modular skid mounted engineering design, prefabrication at the engineering site, and multiple implementations of this modular design. For biofuels from local biomass, such as saw dust, this appears to be commercially attractive.

Bibliography

[1]   Baldea M, Edgar TF, Stanley BL, Kiss AA. Modular manufacturing processes: Status, challenges, and opportunities. AIChE Journal. 2017 Oct;63(10):4262–72.

[2]   Patience GS, Boffito DC. Distributed production: Scale-up vs experience. Journal of Advanced Manufacturing and Processing. 2020 Apr;2(2):e10039.

[3]   Weber RS, Snowden-Swan LJ. The economics of numbering up a chemical process enterprise. Journal of Advanced Manufacturing and Processing. 2019 Apr;1(1–2):e10011.

[4]   Bieringer T, Bramslepe C, Brand S, Brodhagen A, Dreiser C, Fleischer-Trebes C, Kockmann N. Modular plants: Flexible chemical production by modularization and standardization-status quo and future trends. DECHEMA Gesellschaft für Chemische Technik und Biotechnologie eV; 2016.

[5]   Othman MR, Repke JU, Wozny G, Huang Y. A modular approach to sustainability assessment and decision support in chemical process design. Industrial & Engineering Chemistry Research. 2010 Sep 1;49(17):7870–81.

[6]   O'Connor JT, Kowall CP, Haapala KR, Agrawal NV, Paul BK. Specialty chemicals production case study: Economic analysis of modular chemical process intensification versus conventional stick-built approaches. Journal of Advanced Manufacturing and Processing. 2021 Oct;3(4):e10102.

[7]   Sievers S, Seifert T, Schembecker G, Bramsiepe C. Methodology for evaluating modular production concepts. Chemical Engineering Science. 2016 Nov 22;155:153–66.

[8]   Steyn J, van Heerden F. Modularisation in the Process Industry 2015 Sourced 1 Oct. 2024. https://www.ownerteamconsult.com/wp-content/uploads/2020/03/Insight-Article-019-Modularisation.pdf.

[9]   LePree J. Moving to modular. Chemical Engineering. 2015 Jan 1;122(1):16.

[10] Halfort S, Kretzschmar T. Practical Solutions to Modular Project Execution, Fluor Corp. June 1, 2017 | Sourced 13 March 2023. https://www.chemengonline.com/practical-solutions-modular-project-execution/.

[11] Harmsen J, Verkerk M. Process Intensification: Breakthrough in Design, Industrial Innovation Practices, and Education. Walter de Gruyter GmbH & Co KG; 2020 Jul 20.

[12] BTG bioliquids. 2023. Sourced 2—3-2023: https://www.btg-bioliquids.com/plants/.

4 Design for energy efficiency

4.1 Basics design for energy efficiency

Processes and the laws of thermodynamics

Chemical processes in general lose little heat by conduction to their local environment because all hot equipment such as reactors and distillation columns are well insulated. This insulation ensures that the performance of the processes inside the equipment occur as intended by the designer. Due to this insulation, process energy is not lost to the environment. This means that the first law of thermodynamics—energy is conserved— and can be used for all energy flows inside the process. So, these energy flows can be reused.

However, the second law of thermodynamics states that energy can only flow from a high temperature to a low temperature. The second law is about the quality of energy. For heat, this quality is given by the temperature of the heat. The quality of any energy in any form is given by the quality factor q_1. This is a factor for energy flow entering the object system (the process). Energy entering a system times the q factor is called exergy. So, exergy is the useful energy of a stream.

For heat this quality factor, called q_1, is provided below, obtained from [1]:

$$q_1 = 1 - \frac{T_0}{(T_1 - T_0)} \ln\left(\frac{T_1}{T_0}\right) \tag{4.1}$$

Table 4.1 shows q_1 values for some energy types relevant for chemical engineering [1].

Table 4.1: Energy quality factor q_1 for some energy types [1].

Energy type	q_1
Solar radiation	1
Electricity	1
Exothermic chemical reaction	± 1
Hot combustion gas	0.6–0.8
Chimney gas	0.2
Water 100 °C	0.12
Water 50 °C	0.06
Water 15 °C	0.00

Due to the first law of thermodynamics, energy from hot streams can be used to heat up cold streams. The restriction is given by the second law, that heat only flows from a high temperature to a lower temperature. Based on these two laws, methods have been developed to maximize the use of available high temperature of streams to heat up cold streams. This design field is called heat integration, also called design

https://doi.org/10.1515/9783111203256-006

for energy efficiency. It is also treated in heat exchanger network designs. The next section describes process design for energy efficiency based on the first and second law of thermodynamics.

4.2 Process design for energy efficiency method

4.2.1 Heat exchanger network design using grant composite curve

The starting point of process design for energy efficiency is a process flow scheme with all unit operations. For each unit operation and subsections, such as the distillation re-boiler and condenser, enthalpy flows and operating temperatures should be listed. With this information, a grand composite curve is made. Guidelines are provided in textbooks by Douglas [2], Seider [3], Smith [4]. An example is shown in Figure 4.1, obtained from [5]. The vertical axis is the temperature, and the horizontal axis is the enthalpy flow. All unit operations that produce energy are placed in the hot composite curve with a sequence of increasing temperature. All unit operations that require energy are placed in the cold composite curve also with increasing temperature.

The location of the smallest temperature difference between the hot composite curve and the cold composite curve is called the pinch. This analysis method is called pinch analysis. The additional heat required Q_h and the net heat to be cooled Q_c can both be directly found from this grand composite curve. The amount of heat that can be recovered $Q_{recovery}$ is also shown.

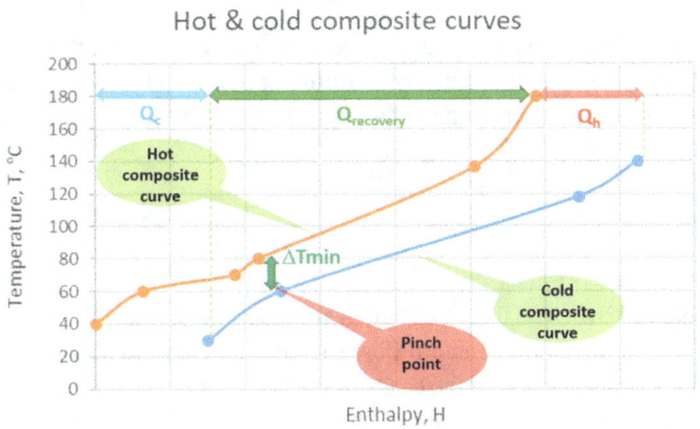

Figure 4.1: Grand composite curve obtained from [5].

From this composite curve, a heat exchanger network can be designed to use all available heat of the hot composite curve for heating up unit operations of the cold composite curve, by designing a network of heat exchangers.

This heat exchanger network is to be designed by these rules [2].
Rule 1: Do not transfer heat across the pinch
Rule 2: Add heat only above the pinch
Rule 3: Cool only below the pinch

By placing heat exchangers in the process using these rules, the additional heat supplied to the process is greatly reduced. Energy savings of 30–50 % are easily obtained [2]. This method of placing heat exchangers to reuse energy is called process heat integration design. It can also be used for existing processes.

4.2.2 Temperature choice exothermic reaction

The composite curve shows that it is advantageous to have the heat production of exothermic reactions at a temperature as high as is feasible. In general, the reaction temperature is chosen such as to minimize by-product (waste) formation. However, often there is still some freedom to choose the reaction temperature.

In pharmaceuticals, often the chemist chooses a low reaction temperature to facilitate easy experimentation in the laboratory. The green chemistry principle on energy requirements by Anastas [6] even states that synthetic methods should be applied at ambient temperature. For exothermic reactions, and most chemical reactions are exothermic; this is a wrong principle. As all reaction heat produced is produced at an ambient temperature, it is useless. Even worse, this heat has to be removed for temperature control, and due to the ambient temperature, this requires a heat pump to remove it; hence, it will be expensive.

So, exothermic reactions need to be run at higher than ambient temperature. If there are no side reactions involved, then the temperature should be chosen as high as is practically feasible, so that most reaction heat energy can be used for heat integration.

4.2.3 Software for heat integration

Nowadays software programs are available to design heat integration by heat exchangers. Often these programs can be combined with flow sheeter programs. Heat integration over large parts of the process can then be easily made.

4.3 Example case: Exothermic reactor and heat integration design

A theoretical case of an exothermic reactor and heat integration design is shown in Figure 4.2 [7]. It shows a reactor with an exothermic reaction and two heat exchangers. Heat exchanger HX1 heats up the feed stream S1 to the reactor inlet temperature using a heat source. Heat exchanger HX2 cools the reactor output stream to near ambient temperature. The heating requirement is 500 kW for this case. It is clear in this design that the heat of the reaction is just cooled away to 40 °C, and hence is not used.

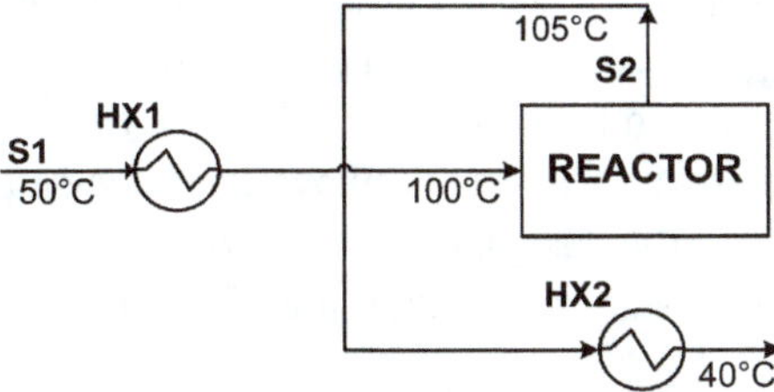

Figure 4.2: Heat exchanger design with no heat integration.

In the heat integrated design shown in Figure 4.3, heat exchanger HX 1 now uses the reaction heat to heat up the feed stream. HX3 further heats up the feed stream to the required temperature of 100 °C. Now the heating requirement is only 140 kW, so over 70 % energy is saved by this heat exchanger design.

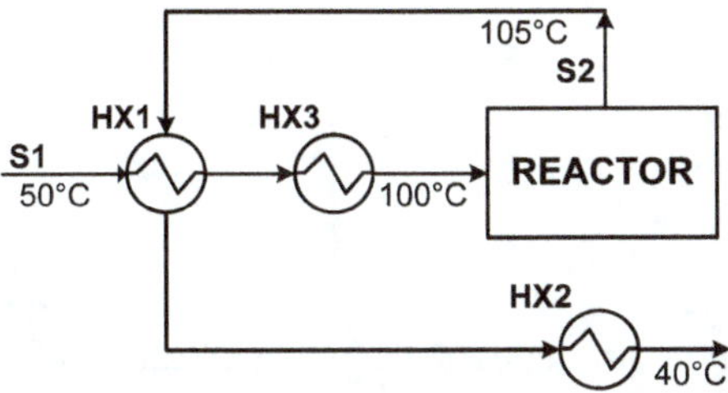

Figure 4.3: Heat exchanger design for energy savings.

Bibliography

[1] De Swaan Arons J, van der Kooi H, Sankaranarayanan K. Efficiency and Sustainability in the Energy and Chemical Industries. New York: Marcel Dekker Inc.; 2004.
[2] Douglas JM. Conceptual Design of Chemical Processes. New York: McGraw-Hill; 1988.
[3] Seider WD, Seader JD, Lewin DR Product & Process Design Principles: Synthesis, Analysis and Evaluation. Hoboken: John Wiley & Sons; 2009.
[4] Smith R. Chemical Process Design. New York: McGraw-Hill; 1995.
[5] Guidebook for Successful Development of Process Simulation Projects. Sourced 02-10-2024. https://simulatelive.com/learn/lessons/how-composite-curves-and-pinch-analysis-are-used-to-improve-the-energy-efficiency-of-the-plant.
[6] Anastas P, Eghbali N. Green chemistry: Principles and practice. Chemical Society Reviews. 2010;39(1):301–12.
[7] Harmsen J, et al. Product and Process Design—Driving Sustainable Innovation. Berlin: De Gruyter; 2024.

5 Evaluating process designs

5.1 Basics process design evaluation

Evaluating a process design is an important step of the design cycle. In that step, the design result is compared with the design goal, scope, and criteria. If the design does not meet the goal, or any other of the criteria, the design has to be modified until it meets all criteria. If the design is made inside a company with a formal innovation stage-gate system, then a stage-gate panel will also evaluate the design. For each next stage-gate, the number of criteria and the accuracy of the design information for the evaluation increases.

Our main focus in this chapter is on providing guidelines for evaluation criteria for the process concept stage. However, at the end of this chapter, a criteria set is provided for designs in the final innovation stages. This criteria set is also suitable for designs with strong connections with society, such as urban-industrial complex designs.

Evaluating a process concept design by comparing it to the best existing process for the same product that is in commercial operation helps enormously. The purpose of a new process concept design is after all to obtain a better process than the best existing process.

5.2 Evaluation method

5.2.1 Process safety evaluation

The first safety evaluation method is based on the strategies of achieving inherent safer designs stated in Section 2.2. These can be used in process design evaluations by taking the best process in commercial scale operation as a reference process and then compare the concept design with that reference process. To this end, the strategies are changed into questions and strategy, number 3 is ignored as it cannot simply changed into a clear question. Here are then the three questions to be used for a safety evaluation of a new design.

1. Is the quantity of material, energy, or energy density reduced compared to the reference?
2. Is hazardous material compared to the reference process reduced or eliminated?
3. Has the new process less process steps and equipment?

The second safety evaluation method is the Dow Fire & Explosion Index (DFEI). It is a practical method to evaluate process designs in the concept stage as only input information is needed about the chemicals involved, their vapor-liquid properties, and the temperatures and pressures in the process, to determine the DFEI value [1]. If the value is above 100, then the process design is considered unsafe. A redesign is then needed

https://doi.org/10.1515/9783111203256-007

with alternative chemicals and/or process temperatures. The method is described for process industries in a practical way by Hemmer [2].

In the EPC, innovation stage safety risks are assessed in detail and mitigation measures are included in the detailed design if the risks are deemed to be too high [3].

5.2.2 Environmental impact by rapid life cycle assessment

Life cycle assessment is the established method for assessing the environmental impact of any design [4]. The method has, however, been developed for product design. Even simplified methods for product concept design still require a large effort and are therefore not suitable for process concept design.

Therefore, Jonker developed a rapid LCA method for process design. It follows all steps of the generic LCA method but requires far less information [5]. It is also far less accurate than a full LCA, but in process concept design that lack of accuracy is not so much of a problem, as the new process should just have far a lower environmental impact than the best existing commercial scale process for the same product. That process is called the reference case. The LCA steps and the impact types after the new process design will be the same as for the reference case, hence this simplification has a sound base. The rapid LCA is just used to evaluate whether the new design has indeed this lower environmental impact when compared to the commercial scale reference case.

The rapid LCA method of Jonker [5] consists of the following eight steps:

1: Define functional unit
2: Define life-cycle study goal
3: Define life-cycle scope
4: Select one or two key environmental impact types
5: Identify all major emission streams for each life-cycle step
6: Quantify the environmental impact
7: Compare novel design with the reference case
8: Conclude whether the novel design has a lower environmental impact or not.

Step 1: Define Functional Unit (F. U.)

The functional unit states what performance is needed for the product design. For a process design, this is simply the description of the existing commercial product to be produced by the process to be designed. If the process reaction produces two products, then the functional unit should be defined for the two products. If, for instance, the two products are styrene and propylene oxide, the functional unit could be defined as 1 kg styrene and 1 kg propylene oxide.

If it is desired to know the individual environmental impact of each of the two products, then the emissions have to be allocated in some way to each of the two products.

This is the so-called allocation problem for LCA [6]. There are various ways to solve the allocation problem.

As the two products are coupled by the reaction and the whole process is designed for these two coupled products, the simplest way is to use the mass ratio of the two products and allocate the emissions to each product by this ratio.

Step 2: Define life-cycle assessment goal

For any LCA, the goal of the LCA has to be set. For a process concept design, an environmental impact assessment of the design (or alternative designs) in comparison with the established commercial process in such a way that it can result in a conclusion that the new design has a higher or lower environmental impact compared to the established commercial process.

Step 3: Define life-cycle scope

The scope definition contains the main life-cycle boundaries. These boundaries can be chosen with the help of Figure 5.1.

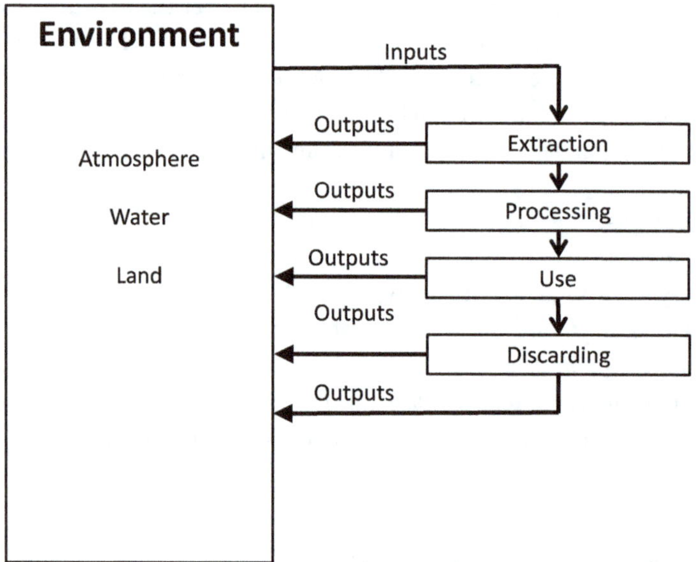

Figure 5.1: Life cycle assessment system elements.

In general, the first life-cycle step is to extract material from the environment (cradle), which after transformations enters the product. The last step is often the disposal of the product after its use (grave), or its upgrade step from waste to a reused product.

Many more life-cycle steps can be considered, for instance, the steps leading to the process equipment for mining and further processing. For process concept design, these steps can be neglected, as for most cases these contributions to the environmental emissions are small compared to the emissions in the main life-cycle string. The reason that these contributions are relatively small is due to the fact that the amount of process construction material is, but a small fraction of the total product mass produced over its entire lifetime. A process of 100,000 t/year production capacity typically has a lifetime of over 10 years so 1,000,000 tons of product are produced over its lifetime. The total amount of construction materials is in order of 1,000 tons. So, the mass of construction material is 0.1 % of the product mass produced of the process lifetime. Impact factors (effect on environment per amount of product) are in general 1–10. Hence, the environmental impact of the construction material is less than 1 % of the total environmental impact.

This reasoning only holds for cases for which at the end of the process life is properly dismantled without causing environmental impact. For nuclear power stations, for instance, this reasoning does not hold, as the construction material at the end of the process lifetime is radioactive and dismantling causes serious environmental impact in general.

Transportation in between some main life-cycle steps can be included in the life cycle. This is relevant if carbon dioxide emissions of transportation are significant compared to other carbon dioxide emissions. For bulk products transported by ship, this contribution to the overall carbon dioxide will be small and can be neglected in the concept stage of the design.

Life cycles are best represented by block-flow diagrams in which each block represents a process step, and each arrow represents a mass flow. Standard methods for mass balance checks can be applied to ensure the quality of the life cycle.

Step 4: Select one or two key environmental impact types

Table 5.1 shows all environmental impact types of a life-cycle assessment, obtained from Hauser [4]. For a concept design, it is not necessary and also not desired to do a full life-cycle assessment for all impact types. A literature search on the product for environ-

Table 5.1: Environmental impact types for life-cycle assessment [4].

Atmosphere	Water	Biotic	Soil
Global warming	Ecotoxicity	Ecotoxicity	Ecotoxicity
Ozone layer depletion	Nitrification	Habitat reduction	Physical degradation
Photochemical pollutants	Heat	Biotic resource reduction	Dehydration
Acidification		Physical ecosystem reduction	Resource depletion
Radiation			
Noise			
Smell			
Visual landscape			

mental impact can be executed to find the types of impact that are most concern. If no information is available about the product, then global warming may be chosen as the main impact type. As global warming is a concern for most process designs.

Step 5: Identify all major emission streams for each life-cycle step
The best way to identify all major inputs from and emissions to the environment is to execute a literature search on the life cycle of the product. If that is not available, then search for an LCA of a product made from the product may give the required information.

Step 6: Quantify the emission streams and the resulting environmental impacts
The emissions to the environment should be quantified. The emissions should also be related to the function unit dimension chosen. If the function is, for instance, expressed as 1 kg, then the emissions should be quantified relative to the functional unit. Often the emissions can then be expressed as a dimensionless unit, for instance, in the case of carbon dioxide emissions as kg CO_2/kg product. As soon as the production capacity of the product is defined, the total emissions for each emission type can also be defined.

Quality check on emission calculations
The quality of the emissions calculated can be quickly evaluated by a simple atom balance over the process input and output streams. Any atom that enters the process will leave the process. Often when a crude input such as biomass is used as input to the process the concept designer neglects or forgets that the biomass contains atoms such as potassium and sulphur, which will leave the process in one of the outlet streams. A simple atom balance analysis then reveals that the output streams are not sufficiently defined, and that the envisaged output destinations of some output streams have become infeasible.

Pitfall in emission dimensions
A pitfall in emissions is the wrong dimension obtained from some publications. Economists and also policymakers often express emission amounts in absolute mass dimension (ton), while they mean the amount per year (ton/year). So, check what is meant and also do not copy this habit of expressing the mass flow in mass. Eventually errors will be made further down the road of using the report.

Step 7: Compare novel design with the reference case
If more than one impact type is used in the comparison between the new design and the reference case and for one impact type, the new design has a lower impact and for the

other impact type it has a higher impact. Then a conclusion on the new design cannot be simply drawn.

In LCA literature, this problem is called valuation [4]. The problem is solved in textbooks on LCA by using weight factors for each impact type so that the results of the two impact types can be added up to a single number.

However, the weight factors are subjective and depend on the world view of the person or group of persons that make the weight factors. Also, inside a company the discussion on whether the new design is better or not on environmental impact will often be endless.

It is far better to consider the new design and modify it in such a way that it will have lower impacts for both types of impact. Such a design is then robust to different opinions inside and outside the company and to the opinions of various stakeholders on the relative importance of impact types. Also, for marketing purposes the novel product will then be conceived as far better.

A helpful method for a rapid evaluation in the concept stage is to fill in template Table 5.2 for all environmental impacts for which data are available. For impacts with no data "no data yet" is filled in. For some of those impacts, it may be possible to reason that the impact for modern design will be lower than the reference case. Then "lower" can be filled in the fourth column.

Table 5.2: Template environmental impact ratio Innovative design versus reference case.

Impact type	Reference impact	Innovative design impact	Impact Ratio new/reference

The numbers of the last column are used for the evaluation. If all ratios are less than 1, then the environmental evaluation has a positive outcome. If for some impacts the ratio is less than 1, but for at least one impact the ratio is higher, then the outcome is negative.

Step 8: Conclude whether the novel design has a lower environmental impact or not
Finally, a conclusion should be drawn on the environmental impacts of the new process design compared to the reference case, because the rapid LCA method is very inaccurate. The calculated emissions can be 50 % lower or 100 % higher. Firm conclusions can only be drawn if the design has at least a factor 2 lower emissions to the environment compared to the existing process. For global warming, the impact should be near zero, as the design will be implemented several years after the concept design and will then be in operation beyond the year 2050, when processes should have a net zero impact on global warming; see Section 5.2.5 on Sustainable Development Evaluation including Global Warming Impact.

5.2.3 Technical Readiness Level evaluation

A Technical Readiness Level (TRL) evaluation is an objective method for determining the maturity of a technology. It is based on nine well-defined levels by which its maturity can be established. It can be used for buying novel technology under development to find out what the maturity status of that technology is.

For start-up companies looking for funding of the next step in developing the new process, TRL determinations are useful to obtain funding for reaching next TRL levels and to sell their technology, as TRL terms are known by subsidy providers.

Table 5.3 shows the definitions of NASA and the European Union TRL. There is an important difference between these two. The NASA definition includes "reported." This is a very important addition, as the TRL should be based on written reports. NASA also includes "application formulated." This is also an important addition, because the maturity of technology can only be determined by its specific application.

Table 5.3: TRL Definitions according to NASA [7] and the European Union [8].

TRL	NASA usage [7]	European Union [8]
1	Basic principles observed and reported	Basic principles observed
2	Technology concept and/or application formulated	Technology concept formulated
3	Analytical and experimental critical function and/or characteristic proof-of concept	Experimental proof of concept
4	Component and/or breadboard validation in laboratory environment	Technology validated in lab
5	Component and/or breadboard validation in relevant environment	Technology validated in industrial environment in the case of key enabling technologies
6	System/subsystem model or prototype demonstration in a relevant environment (ground or space)	Technology demonstrated in relevant environment (industrially relevant environment in the case of key enabling technologies)
7	System prototype demonstration in a space environment	System prototype demonstration in operational environment
8	Actual system completed and "flight qualified" through test and demonstration (ground or space)	System complete and qualified
9	Actual system "flight proven" through successful mission operations	Actual system proven in operational environment (competitive manufacturing in the case of key enabling technologies, or in space)

For applying TRL for processes, Table 5.4 is provided, with terms used in process technology innovations. This table is derived from the NASA and EU TRL tables. The author added TRL 0 so that ideas can be discussed and assessed that have not reached TRL 1.

Table 5.4: Technology readiness level definitions for novel processes derived from NASA and EU TRL definitions.

TRL level	Description
	All information must be in written reports
0	Idea of novel process
1	Experimental proof of key individual novel principles of process
2	Concept design of process
3	Experimental proof of process concept
4	Experimental validation of integrated process (any scale)
	Process Industry Environment
5	Industrial professional's technoeconomics assessment product and process
6	Novel product prototype tested by industry
	Novel process technology demonstrated by industrial pilot plant
	Commercial Production Environment
7	Novel process launched at commercial scale and in steady state operation
8	Learning points first plant incorporated in commercial process designs
9	Commercial processes in operation and evaluation reports available

In my 14 years of consultancy for various companies in process industry branches, I have heard several misconceptions in the use of TRL. To avoid these misconceptions, some points of the TRL table are highlighted.

For the concept stage, a TRL value of 2 can be achieved by making process concept design.

If that design is validated by an experimental program executed for all new critical items of the process concept, then a TRL level of 3 can be achieved.

If an experimental program in which all process elements are integrated and is successfully executed, then TRL 4 is obtained.

There is a major shift from TRL 4 to TRL 5. TRL 5 and higher can only be achieved by industrial practitioners. So, statements or reports by academics about having achieved TRL 5 or higher are not valid. TRL 5 is achieved by the industry if the new process elements are experimentally validated.

TRL 6 is achieved if an integrated pilot plant is successfully operated in an industrial company.

There is another major shift from TRL 6 to TRL 7. TRL 7 can only be achieved by having the commercial scale process in a commercial scale operation environment. Hence, demonstration plants at R&D facilities cannot achieve TRL 7.

TRL 8 is achieved if all learning points of the commercial scale operation are used to improve the process design.

TRL 9 is achieved when the improved design is implemented at commercial scale and steady-state operation is achieved.

Warning

TRL evaluations are mostly used for buying and selling new technologies, which are still in the R&D stages. It is then very important that TRL is reliably established. TRL determinations are however hard to be made, and especially hard to be made by R&D practitioners. The author gives courses on process scale-up and these courses contain exercises about determining a TRL from written information. Over 90 % of the course participants arrive at TRL values that are levels 1–2 too high. When asked, they explained that they guessed the TRL from the written information but did not check whether the required information for reaching a TRL was stated in the text. R&D practitioners in general read information and interpret it for the usefulness of their R&D. For determining TRL, however, interpretation should be minimized to what is actually stated in the text. So, the text should be read as lawyers do.

5.2.4 Economical evaluation

The simplest economic evaluation of process designs is using the Economic Potential (EP) approach of Douglas. It starts with the revenue (sales) per year. This is simply the sales price of the product (€/ton) multiplied by the process capacity (ton/year). Then the simplest Economic Potential EP_1 is calculated. To that end, the feed cost (price/ton) times the feed rate (ton/year) is subtracted from the revenue. The cost of carbon dioxide emitted to the atmosphere can also be taken in the same way as feedstock cost and can also be used to determine EP_1.

$$EP_1 = \text{Revenue} - \text{Feed cost (€/y)} \tag{5.1}$$

If $EP_1 < 0$, then the concept design is not profitable. As for conventional products and feeds, the future prices will be affected in a similar way by inflation and other factors this analysis is good enough for a decision to stop the development of this novel design.

If $EP_1 \gg 0$, then the next cost item of importance being the energy cost is subtracted from EP_1

$$EP_2 = EP_1 - \text{Energy cost} \tag{5.2}$$

Again, if $EP_2 < 0$, then the further development of the design can be stopped.

The next item of importance is the investment cost. To get it as a cost per year, the investment is divided by the required payback time (years). To obtain the investment cost figure for the concept design, cost various investment estimate methods are available from textbooks. The simplest method is by Bridgwater. It is based on the number of functional process steps of the concept design. Functional process steps of a process are the steps that fulfill a function. Examples of functional process steps are heating, mixing, reaction, distillation, and packaging. A function consists in general of one unit operation. If, however, several of the same unit operations fulfil a function, for instance, a number of reactors in parallel or in series then that is counted as one functional process step.

An investment cost estimate correlation is provided by Bridgewater.

For process capacities >60 kton/year, the correlation is

$$C_{DPC} = 4.3 \times 10^3 \times N \times (Q/S)^{0.675} \tag{5.3}$$

For process capacities <30 kton/year, the correlation is

$$C_{DPC} = 3.8 \times 10^5 \times N \times (Q/S)^{0.30} \tag{5.4}$$

C_{DPC} = Capital investment of Process Inside battery limits ($)

N = Number of functional process steps

Q = Plant production capacity (ton/year)

S = Reactor single pass conversion fraction of Process Input

The correlation was published by Bridgewater in 1974 [9]. So, an inflation correction factor (Corinf) is needed to get a current estimate.

The investment cost estimate with inflation correction then is

$$C_{DPCinf} = Corinf \times C_{DPC} \tag{5.5}$$

The inflation correction factor from 1974 until 2024 is 5.4 [10]. So, Corinf = 5.4.

This capital investment cost estimate holds for process design inside battery limits. This means that outside battery limits such as storage and utility investment cost are not taken into account. This cost estimate can only be used for comparisons between process design alternatives. Absolute accuracy is poor.

Furthermore, the correlation can be used for process concept designs with conventional unit operations. For process designs based on functional integrated process intensifications, the correlation becomes even more unreliable. The process intensified process of Eastman for the production of methyl acetate, for instance, has a factor 5 lower investment cost than the process design based on 11 unit operations; while using the Bridgewater correlation the investment cost would have been a factor 11 lower.

A remark should be made on the use of the term functional process step. This means a process step characterized by a function. Functions are, for instance, heating, mixing, reaction, distillation, and packaging. A process function can contain more than one of the same unit operation. For instance, a series of reactors to produce the product is a single functional reaction process step.

Also for nonconventional process steps such as, for instance, electrolyzers that cannot be used. Other sources of information should then be used, which are outside the scope of this book.

For a payback time of 3 years, the annual investment cost (AIC) is then given by

$$AIC = C_{DPC}/3 \tag{5.6}$$

Most large companies have a yardstick for the payback time of a novel technology in the concept stage of 3 years. This background reasoning for this short payback time is that in the further development of the process concept more cost items will emerge. With this annual investment cost, EP_3 is calculated.

$$EP_3 = EP_2 - \text{Annual Investment cost} \tag{5.7}$$

If $EP_3 < 0$ then the concept design development can be stopped.

Other costs, such as maintenance and operational personnel costs, are in general less than 10 % of the annual investment cost. So, 10 % is assumed for the first economic evaluation. So, EP_4 can be defined as

$$EP_4 = EP_3 - 0.1\,AIC \tag{5.8}$$

5.2.5 Sustainable evaluation

Sustainable evaluation is about evaluating whether the design contributes to the Sustainable Development Goals (SDG) of the United Nations [11]. These Goals are subdivided into 5 sections People (Social), Planet (environment) Prosperity (Social), Partnership, and Peace. The first three are called Triple P and are seen as a minimum for companies to state that they contribute to sustainable development. It is then called: 'Triple Bottom Line' For a concept design the SDG's are too qualitative. However the SDG's have a deeper level containing targets and these are quantitative and can be used for evaluating designs. By scanning the Triple P SDG's a selection on relevant selection of at least one SDG from each P. and then selecting at least one target per SDG and useful criteria set is obtained.

United Nations Sustainable Development Goals for 2030 [11].
People (Social) Goals
Goal 1. End poverty in all its forms everywhere
Goal 2. End hunger, achieve food security. and improved nutrition and promote sustainable agriculture
Goal 3. Ensure healthy lives and promote well-being for all ages
Goal 4. Ensure inclusive and equitable quality education and promote lifelong learning opportunities for all
Goal 5. Achieve gender equality and empower all women and girls
Prosperity Goals
Goal 6. Ensure availability and sustainable management of water and sanitation for all
Goal 7 Ensure access to affordable, reliable, sustainable, and modern energy for all
Goal 8. Promote sustained, inclusive, and sustainable economic growth, full and productive employment, and decent work for all
Goal 9. Build resilient infrastructure, promote inclusive and sustainable industrialization and foster innovation
Goal 10. Reduce inequality within and among countries
Environmental (Planet) Goals
Goal 11. Make cities and human settlements inclusive, safe, resilient, and sustainable
Goal 12. Ensure sustainable consumption and production patterns
Goal 13. Take urgent action to combat climate change and its impacts*
Goal 14. Conserve and sustainably use the oceans, seas, and marine resources for sustainable development
Goal 15. Protect, restore, and promote sustainable use of terrestrial ecosystems, sustainably manage forests, combat desertification, and halt and reverse land degradation and halt biodiversity loss
Peace Goal
Goal 16. Promote peaceful and inclusive societies for sustainable development, provide access to justice for all, and build effective, accountable, and inclusive institutions at all levels
Partnership Goal
Goal 17. Strengthen the means of implementation and revitalize the Global Partnership for Sustainable Development

For the environmental goal, also the target set to combat climate change by the UN meeting COP21 in Paris 2015 should be added [12]. The Paris Agreement states that global warming should stay well below 2 °C and preferably below 1.5 °C. It is up to the country governments, or the EU, to define measures to reach that target.

The EU aims to be climate-neutral by 2050—an economy with net-zero greenhouse gas emissions. This objective is at the heart of the European Green Deal and is a legally binding target thanks to the European Climate Law [13]. Other countries may also make such a legal binding target.

For process concept design, this means that the concept design should be such that it has a net-zero emission when taken in operation, as it is likely that the process will be in operation in 2050 and beyond. It also means that the design has to have a net-zero emission over its life cycle. As the global warming gas emissions associated with the construction material, steel and concrete are very small over their life cycle; see the LCA Section 5.2.2. For the product produced by the process, the net zero(connect to next line here)emission also has to be obtained. This subject is however beyond the scope of this book.

5.2.6 Evaluating design fit to reality

A design is most successfully implemented at commercial scale when it has a good fit with reality in all aspects. To make a design fit to reality in all aspects, a good description of reality is then needed. This is of course an enormous task and, moreover, cannot be done for a concept design, as the local and global reality where the final design is implemented is still far away in the future. However, an attempt can be made by using a model of reality created by the Dutch philosopher Dooyeweerd and described by Verkerk for design purposes [14]. From this description, design criteria for fitness to reality have been derived by Harmsen [15]. The criteria with examples are described in Table 5.5 below.

Table 5.5: Modal aspects for design criteria.

Modal aspect	Criteria type	Examples
Belief	Agreement with belief	Halal, Kosher
Moral	Moral values	Safe, Health, Sustainable
Juridical	Legal	Within the laws
Esthetical	Esthetic	Attractive appearance
Economic	Economic	Profitable, efficient
Social	Social	Socially accepted
Linguistic	Communication	Clear reporting
Shaping	Shape	Correct 3-D shape
Logical	Logical	Correct model calculations
Sensitive	Sense	Tasty product
Biotic	Biological	No patogen contamination
Physical/Chemical	Properties and principles	Technically feasible
Kinematical	Flow rates	Heat and mass flows correct
Spatial	Geometry	Within plot area
Numerical	Numerical	Mass balances closed

Regarding the modal aspect "belief"; design criteria for this aspect can be generated by first considering the majority belief of the country or countries for which the design is relevant. If it is about product design and the majority belief of the country is Islamic, then a design criterion can be that the product is to be halal. For food, this is obvious, but other personal care products such as soap, also halal can be desired. By inviting a religious person, an official document may be obtained that the product has been halal approved by the official person. If the design is a process and as a solvent ethanol is considered, then a religious person can be invited to obtain advice about which alternative solvent may be acceptable. Also, for other beliefs feed with no-animal origin can be a criterion.

Also, the choice of the process location can be subject of belief aspects. A process for Liquefied Natural Gas (LNG) was planned for a location in the north part of Australia. Later in the project, when its location was publicized, it appeared that the location was considered holy ground by the Aboriginal inhabitants. The location had to be changed due to their protests. As a consequence, the design also had to be modified, due to the different conditions at the new location. So, considering the modal aspect belief and deriving criteria from this aspect, can avoid major marketing and other project disasters.

Moral values play a strong role in safety, health, environment, social, and sustainable development criteria. Most companies have developed specific design criteria for these aspects, and report on these in their annual sustainability reports.

Social acceptance of new products and processes ais, of course, of enormous importance. If a product is not accepted, there will be no market for it. Social acceptance is difficult to predict. In most cases, it needs testing.

The linguistic model aspect is about human communication. A design in itself is information to be communicated to others. Therefore, the linguistic model aspect is particularly important. Clear reporting, presenting, and obtaining feedback is vital for any design.

The shaping aspect of design is about the 3-dimensional shape of product and process. The shape of a product is important for its use, transport, and storage. The shape/outline of the process must fit within the given plot limits.

The logical aspect is about reasoning, modeling, and calculations regarding the design. Mass and heat balances, for instance, must be correct.

The sense aspect is particularly important for product design. In most design cases, for each sense at least one criterion will need to be defined.

The biotic aspect is particularly important for biological and food products and processes. Contamination by pathogenic microorganisms, for instance, must be avoided. Also, for process design, biotic criteria are important, for example, prevent microorganism-induced corrosion.

Physical and chemical aspects concern all product and process designs. The technical feasibility of a product or a process will always depend on the physical and chemical properties of the materials selected.

The kinematic aspect is about movement. It is particularly important for flows inside the process. If the velocity becomes higher than the speed of sound, unwanted effects may occur.

The spatial aspect is the space taken up by the product and process, and by the material transport related to the product and process design.

The numerical aspect is involved in all computations made for the design. The calculations should be based on correct mathematical equations.

5.3 Example case: Methanol from ocean-dissolved carbon dioxide

5.3.1 Methanol process concept design and reference process

To illustrate how to evaluate a process concept design with criteria in Section 5.2, a process concept design by Engineering Doctorate students at TU Delft is used as a starting point. It is a methanol-process design using Ocean-Dissolved CO2 as feedstock and sunlight as energy source. The process design and the solar cells are situated on a floating platform. The process design capacity is 100 kt/year of methanol. Methanol is envisaged as a liquid transport fuel replacing fossil petrol. The process concept design is reported as part of the Engineering Doctorate education program [16] and summarized in a publication [17]. The author of this book improved the concept design description and the evaluations.

The process scheme is shown in Figure 5.2. Ocean water is used as a feed for carbon dioxide. Carbon dioxide is obtained by membrane filtration. The ocean water is also used for electrolysis. First, the ocean water is desalinated by membrane filtration. Electrolysis produces hydrogen and oxygen. The latter is not used by send to the atmosphere. (please review the highlighted area for accuracy)

The hydrogenation of carbon dioxide to methanol is carried out in a heterogeneously catalyzed reactor.

The reaction equation and reaction enthalpy of the process design is

$$CO_2 + 6\,H_2 = CH_3OH + H_2O \quad \Delta H298\,K = -49.2\,kJ/mole$$

So, the reaction is exothermic. The reaction heat can be used in the distillation to separate methanol from water.

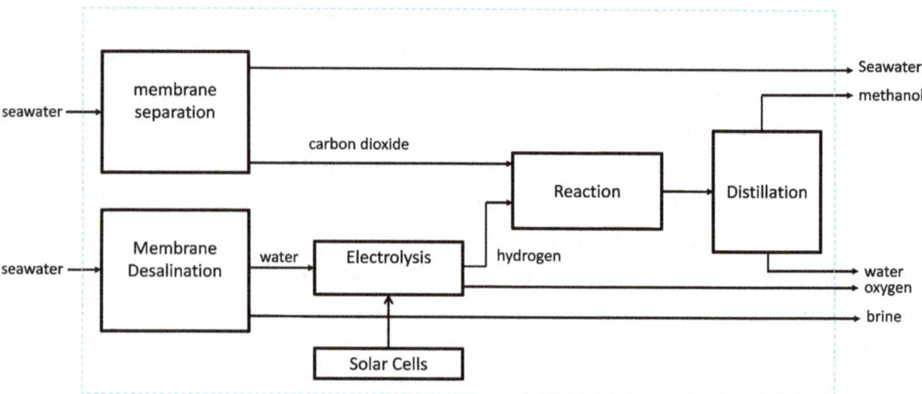

Figure 5.2: Process concept design methanol from carbon dioxide in seawater.

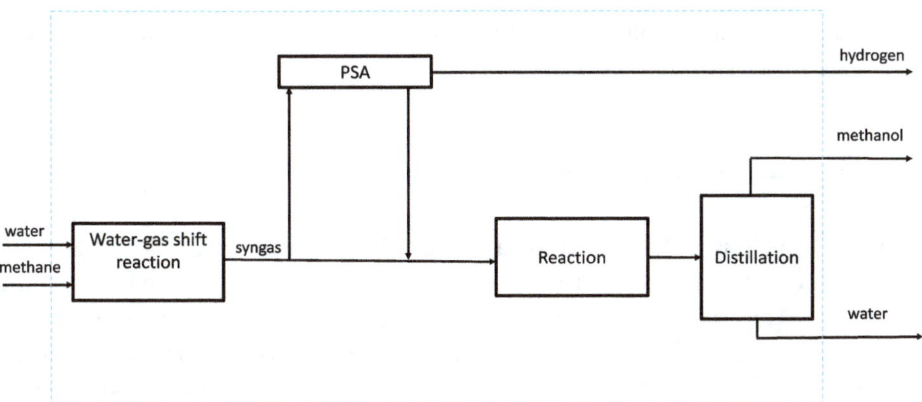

Figure 5.3: Reference case methanol from fossil methane.

Figure 5.3 shows the process scheme of the reference process; the production of methanol from fossil natural gas. The reference process is in operation at commercial scale [18]. For a direct comparison, the commercial scale process inputs and outputs are scaled down to 100 kt/year production of methanol.

The reaction equations and reaction enthalpies of the reference process are given here. The first reaction is steam reforming of methane:

$$CH_4 + H_2O = CO + 3\,H_2 \quad \Delta H298 = 206\,kJ/mol$$

This reaction is highly endothermic and needs heat. This heat is supplied by a furnace where the catalytic tubular reactor is placed (in the furnace, methane is combusted to carbon dioxide and water).

The reaction to methanol for the reference process is

$$CO + 2\,H_2 = CH_3OH \quad \Delta H298\,K = -90.8\,kJ/mole$$

This reaction is exothermic. The CO and H_2 are obtained from methane by steam methane reforming.

The stoichiometric equations show that there is a surplus of hydrogen. This surplus of hydrogen is separated by pressure swing adsorption (psa) from the syngas stream. This surplus hydrogen may go to the fuel pool in a refinery nearby.

Alternatively, as the water gas shift reaction is endothermic and operated at high temperature and carried out inside a furnace, where methane is combusted to provide the heat for the endothermic reaction, Hawkins proposed to use the carbon dioxide of the effluent gas to produce more methanol by hydrogenating this carbon dioxide with the surplus of hydrogen [19]. So, methanol processes from fossil methane may become more efficient than the present conventional process taken as a reference. The overall carbon dioxide emitted over the life cycle of methanol from fossil methane will still be at least 138 kton/a for a 100 kt/a methanol.

5.3.2 Global warming gas emission evaluation

In this concept design, methanol is to be used as a future liquid transport fuel. So, over the life cycle methanol is combusted to carbon dioxide and water and ends up in the atmosphere. The feed of the process is carbon dioxide dissolved in ocean water and oceans absorb carbon dioxide from the atmosphere. By choosing this feed, the carbon cycle over the life cycle is closed; hence, net zero. Carbon dioxide emissions connected to the manufacturing of the solar panels and also the transport of methanol from the floating platform to the coast by chemical cargo ships appeared to be small compared to the carbon dioxide emitted and absorbed and are therefore neglected in this concept design. In the final design, a full LCA will reveal how much (or how little) global warming gas is still emitted.

In the reference case, fossil methane is used for methanol production and that methanol is also considered as transport fuel, so it is also combusted to carbon dioxide and water. Methanol synthesis from methane also requires additional fuel for driving the process, so it causes additional carbon dioxide emission to the atmosphere. This portion is small compared to carbon and, therefore, neglected. For the reference case, the carbon dioxide emitted to the atmosphere by methanol combustion in the use step is 138 kt/year. This is the main carbon dioxide emission over the life cycle.

For the design case, the carbon dioxide emitted by combustion to the atmosphere is absorbed in the ocean, so for that part the net carbon dioxide over the life cycle is zero. The carbon dioxide emitted by manufacturing the solar cells and the process is small compared to the carbon dioxide emitted by the use step and is therefore omitted. So, the concept design is a near net zero carbon process, while the reference case is a large carbon dioxide emitter. So, by 2050 that process will probably be phased out in view of net-zero emission policy by then in place in many countries.

5.3.3 Social impact evaluation

The students considered solar energy as the power source. They therefore chose a location around the equator. Southeast Asia was the most promising location given future predictions of energy and methanol demand. As an exact location, the coast of Indonesia was chosen based on criteria such as storm intensity, water depth, trade routes, and piracy, and also because the Indonesian industry is presently fossil fuel based. So, their concept design would help Indonesia to move to a sustainable economy. The students also did some evaluation of installing fishing nets underneath the solar cells so that also food is produced for the Indonesian economy. They calculated that the economic potential of that fish production could be 50 M€/year.

5.3.4 Economic evaluation

The economic evaluations of the concept design and the reference case are based on the method provided in Section 5.2 and the calculated economic potentials are summarized in Table 5.6. Methane and methanol prices are obtained from Statista from the period 2020–2024 [20]. Prices of methane and methanol vary, however, enormously over this short period and are also dependent on the production location. The methane price in the Middle East, for instance, is far lower than in Western Europe. Moreover, in future carbon dioxide emission, taxation could be in place, increasing the cost of using fossil natural gas as a feed source of transportation fuel.

Table 5.6: Economic evaluation reference case and design.

		Dimension	Reference case	Design
Methane purchase price €		€/ton	200	–
Methane purchase for 100 kt/y methanol design		M€/y	10	
Methanol sales price world market €/ton		€/ton	300	300
Methanol sales for 100 kt/y design		M€/y	30	30
Oxygen sales price		€/ton	–	0
Investment		M€	220	420
Annual investment cost		M€/y	30	140
Economic potential EP_1	Sales – Feed purchase	M€/y	20	30
EP_2	EP_1 – Energy cost	M€/y	20	30
EP_3	EP_2 – Investment cost	M€/y	–10	–110
EP_4	EP_3 – Maintenance cost	M€/y	–13	–124

For oxygen produced as a by-product of the hydrogen production by the electrolyzer, no value is used, because it is produced at sea at ambient pressure. Compressing and transporting to land could be an expense and, therefore, this option is neglected.

The investment cost of the reference case is estimated by the Bridgewater correlation of Section 5.2, assuming four process steps with no recycle. Because it is a commercial scale proved technology, the payback time is put at 6 years (15 % return on investment).

The investment of the solar cells is estimated from published investment cost per MWh electricity produced in 2020 [21] resulting in an investment of 100 M€. The investment of the electrolyzer is estimated based on the required electricity for hydrogen production and uses a reported investment price [22]. The electrolyzer investment is

170 M€. The investment cost for the two membrane process steps that the students report is 10 M€. For the reactor and separator, the Bridgwater method for N = 2 is applied resulting in 110 M$. For the platform installation, the students report an investment cost of 30 M€ [16]. The total investment is then M$ 420. Because the concept design is a totally new design with a high risk of failure and needing considerable R&D before implementation, a payback time of 3 years is set.

For the economic potential values $EP_2 - EP_4$ energy cost for the reference case and the design are neglected. For the reference case, it is likely that surplus hydrogen is used. For the design case, the primary energy is solar based, and the cost of solar panels is included in the investment cost. For operational costs and maintenance costs, 10 % of the annual investment cost is taken.

The economic evaluation shows that the concept design is worth to be pursued as the economic potential EP_1 is higher than for the reference case. The investment costs are, however, still too high to make it a profitable proposition now. The investment cost of the electrolyzer and the solar panels need to come down by factor 3 to make it profitable or the fossil-based methanol sales price has to increase by global warming gas emission taxation.

The student ideas about also to apply fish nets underneath the solar panels and produce fish food is interesting. But this is outside the realm of chemical engineering and needs knowledge from other industry branches to be evaluated.

5.3.5 Technical readiness evaluation

Lindstrom reports experiments with a process at lab scale containing seawater feed to an electrolyzer system to produce hydrogen and oxygen and a carbon dioxide process that extracts carbon dioxide from sea water [23]. Bos studies the hydrogenation of carbon dioxide, and he uses the same catalyst as in the reference case. The single pass conversion of lower, so the recycle stream after methanol separation is higher [24].

The Engineering Doctorate students made an integrated process concept design for all elements [16, 17]. This means that the combination of the referenced results of Lindstrom and Bos, and the student design work results in TRL 2 for the concept design.

Bibliography

[1]   Dow's Fire and Explosion Index Hazard Classification Guide. 7th ed. American Institute of Chemical Engineers (AIChE). ISBN: 978-0-470-93819-5. October 2010.
[2]   Hemmer G. Dow's Fire and Explosion Index, basic understanding & practical application. In: Purdue Process Safety and Assurance Center Spring 2023 Conference. Lafayette, IN, USA, May 8. 2023.
[3]   Bakker HL, Kleijn JP, editors. Management of Engineering Projects: People Are Key. NAP-The Process Industry Competence Network. 2014.
[4]   Hauschild MZ, Rosenbaum RK, Olsen SI. Life Cycle Assessment. Cham: Springer; 2018. https://doi.org/10.1007/978-3-319-56475-3.

[5]   Jonker G, Harmsen J. Engineering for Sustainability: A Practical Guide for Sustainable Design. Elsevier; 2012.

[6]   Heijungs R, Frischknecht R. A special view on the nature of the allocation problem. The International Journal of Life Cycle Assessment. 1998 Nov;3:321–32.

[7]   Nasa TRL definitions. Sourced 02-04-2024. https://www.nasa.gov/pdf/458490main_TRL_Definitions.pdf.

[8]   Technology readiness levels (TRL). Sourced 11—11.2019 Extract from Part 19 – Commission Decision C(2014)4995 (PDF). ec.europa.eu. 2014.

[9]   Bridgewater AV. The functional unit approach to rapid cost estimation. The Cost Engineer. 1974;13(5).

[10]   CPI inflation correction. Sourced 31-10-2025. https://www.bls.gov/data/inflation_calculator.htm.

[11]   UN sustainable development goals. Sourced 17-10-2024. https://sdgs.un.org/goals.

[12]   Paris Climate Agreement. Sourced 11 Oct. 2024. https://unfccc.int/sites/default/files/english_paris_agreement.pdf.

[13]   EU Climate strategies targets. Sourced 16 Oct. 2024. https://climate.ec.europa.eu/eu-action/climate-strategies-targets/2050-long-term-strategy_en.

[14]   Verkerk M, Hoogland J, Van Der Stoep J, de Vries M. Philosophy of Technology: An Introduction for Technology and Business Students. Routledge; 2015.

[15]   Harmsen J, de Haan AB, Swinkels PL. Product and Process Design: Driving Sustainable Innovation. Walter de Gruyter GmbH & Co KG; 2024.

[16]   Sajan A, Conceicao S, Karpov I, Moreno J, Rajhans A. The sustainable design of renewable methanol synthesis from ocean-dissolved co_2 on a floating platform. Student report. TU Delft; 2022.

[17]   Rajhansa A, Conceição S, Sajanc A, Karpovd I, Moreno J. The conceptual design of the production of valuable chemicals from ocean dissolved CO_2, using solar energy on maritime floating platforms. In: CAPE Forum 2022 From Science to Business. University of Twente, Enschede Campus, The Netherlands. 14–16 September 2022.

[18]   Methanex Investor Presentation, September 2014. Sourced 17-10-2024. https://www.methanex.com/sites/default/files/investor/Methanex%20Presentation%20-%20Sept%202014.pdf.

[19]   Hawkins GB. Methanol Synthesis – Theory and Practice. July 30, 2013. Sourced 29-10-2024. https://www.slideshare.net/slideshow/methanol-synthesis-theory-and-operation/24784501.

[20]   Statista. Methanol and methane spot prices. Sourced 25-10-2024. https://www.statista.com/statistics/1323381/monthly-methanol-spot-prices-worldwide-by-region/.

[21]   Renewable Power Generation Costs in 2020. Sourced 7-11-2024. https://www.irena.org/publications/2021/Jun/Renewable-Power-Costs-in-2020.

[22]   Green Hydrogen Cost Reduction: Scaling up Electrolysers to Meet the 1.5 °C Climate Goal. Sourced 7-11-2024. https://www.enlit.world/hydrogen/irena-hydrogen-is-a-cost-competitive-climate-solution/.

[23]   Dara S, Lindstrom M, English J, Bonakdarpour A, Wetton B, Wilkinson DP. Conversion of saline water and dissolved carbon dioxide into value-added chemicals by electrodialysis. Journal of CO_2 Utilization. 2017 May 1;19:177–84.

[24]   Bos M, Brilman D. A novel condensation reactor for efficient CO_2 to methanol conversion for storage of renewable electric energy. Chemical Engineering Journal. 2015 Oct;278:527–32. https://doi.org/10.1016/j.cej.2014.10.059.

Part II: **Embedded process design**

6 Circular economy by design

6.1 Basics circular economy

The widely accepted definition of circular economy is the one by the Ellen MacArthur Foundation, "a circular economy" is one that is restorative and regenerative by design, and which aims to keep products, components, and materials at their highest utility and value at all times, distinguishing between cycles belonging to the biosphere and cycles belonging to the technosphere as shown in Figure 6.1 [1].

For process designs, it means that all input and output streams should be part of closed material cycles so that there is no waste. This implies that if in a design waste is produced then directly a design solution is to be generated to upgrade that waste to a useful product.

Here are two examples of closing cycles. The closed biosphere cycles can be explained by taking biomass as food. This biomass may be simplified to carbohydrates $[CH_2O]_n$. These carbohydrates are converted inside human bodies by oxidation to carbon dioxide and water ending up in air. Via the air, these components plus oxygen are converted by photosynthesis back into biomass. So, the mass balances are closed. Sunlight energy drives this cycle.

The closed technosphere cycles can be explained by using cellulosic biomass for paper production. Wastepaper can be upgraded to paper pulp and reused. In general, this can be done for a limited number of cycles.

In the end, wastepaper may be burned to carbon dioxide and water. These components then become part of the biocycle.

For other materials, such as metals, recycling will be only feasible via the technosphere cycle.

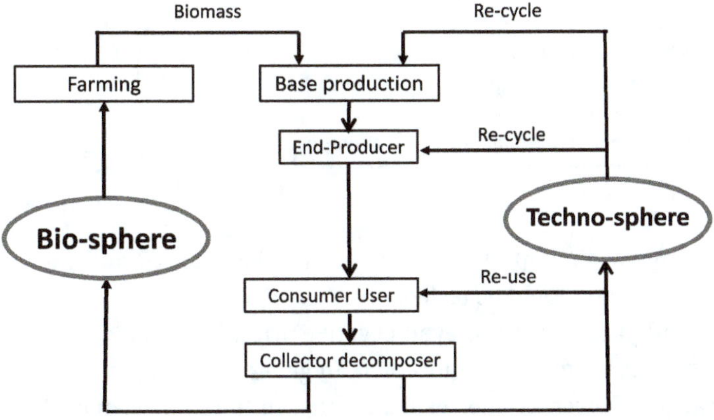

Figure 6.1: Circular economy diagram for closing material cycles. Derived from Ellen MacArthur Foundation [1].

https://doi.org/10.1515/9783111203256-009

For market economies, each entity in the economic cycle has to make a profit. So, value has to be added for each economic partner in the circular economy. This subject is treated in Section 6.2.

6.2 Circular economy by design method

Circular economy cycles are best designed by considering production companies as transforming inputs to outputs. For design purposes, these are then represented by block-flow diagrams. Mass balances are then used to check that no mass is accumulated and also that all input and output streams are accounted for.

Using Economic Potential (EP) calculations with market prices for input and output components, the economic viability of each circular economic cycle can be quickly established. Each partner in the cycle has to have a positive EP value. If not, then the cycle has to change. This can be done by changing the processes. It can also be done by partnerships by which costs are shared such as, for instance, is already happening for large industrial sites where waste upgrading costs are shared by the industrial partners involved [2].

For designing circular economies, Cramer has developed a practical set of 10 guidelines for circular product design. She calls it the Ladder of Circularity [3] (see Table 6.1).

Table 6.1: Ladder of Circularity [3].

Refuse:	Do not use native material but use refused material
Reduce:	Use less material
Renew:	Use waste material in new way
Re-use:	Use waste material again
Repair:	Repair the device or material
Refurbish:	Revive the material
Remanufacture:	Make a new product from used materials
Re-purpose:	Reuse product for other functions
Recycle:	Recycle material for use
Recover:	Recover some value from used material such as energy

The ladder should be started at the top. If a certain step has been tested and cannot be further applied, the lower step should be tested.

As the material used often has a zero or even negative price, while the new product has a higher market price, profits can be foreseen. In general, start-up companies exploit this potential earning power. Cramer estimates that in the Netherlands alone 7 Billion €/year can be made by circular economy, just considering water, energy, and materials [3].

Because the method has economy in its title, it is appropriate to add that circular economy cycles should also be economical. In free market societies, it means that each economic entity in a value chain should make a profit. The author found out that for all design cases explored by EngD trainees at TU Delft, this appeared to be achievable. So, closing material cycles also make economic sense.

6.3 Example case: Solubilized protein from egg-shell waste

6.3.1 Introduction

The industrial production of food ingredients such as mayonnaise is based on chicken eggs as raw material. One of the by-products of this process is Eggshell Membrane (ESM). This material has little value as the protein material is a solid, and insoluble in water or other solvents. A patent by Biova proposes a process to solubilize this ESM to Soluble Eggshell Protein (SEP) [4].

Using this patent as a starting point, students of Technical University of Delft in the Engineering Doctorate program designed a process using circular economy methods described in Section 6.3.2 [5].

6.3.2 Process design

The students started with the Biova process design [4]. They analyzed this process design with the Rapid LCA method, highlighting all environmental emissions. Then they used the ladder of circularity to generate ideas to reduce these emissions. The results are shown in Table 6.2.

Table 6.2: Ladder of circularity applied to eggshell waste to soluble protein.

Refuse:	Eggshells and ESM reframed as valuable by-products
Reduce:	Lower freshwater use by process optimization
Renew:	$CaCO_3$ waste transformed to a new product
Reuse:	Treated wastewater and biogas re\used for energy
Repair:	ESM proteins (e. g., collagen) modified for new uses
Refurbish:	Soluble Egg Shell (SEP) made from eggshell waste
Remanufacture:	SEP and $CaCO_3$ applied in new industry fields
Repurpose:	Found new purposes for eggshell waste, wastewater, biowaste, and $CaCO_3$
Recycle:	Steam/hot water recovered for energy usage
Recover:	Energy generation from biowaste

The students then gathered information about the suppliers of ESM and about potential process industries that could use the soluble protein, and also use all other output

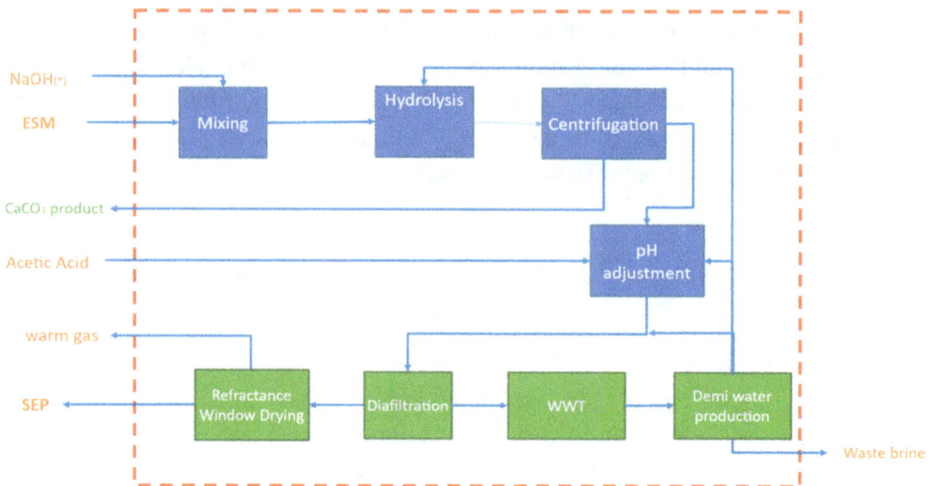

Figure 6.2: Process concept design for soluble eggshell protein.

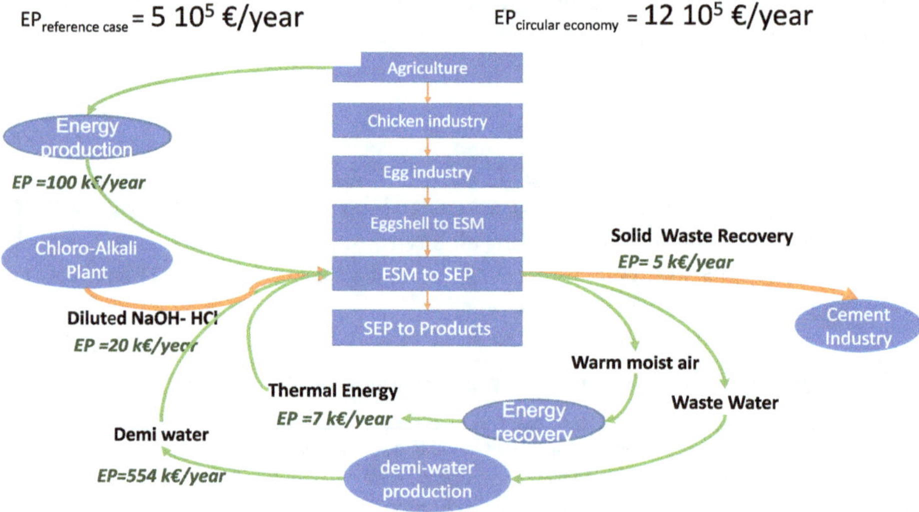

Figure 6.3: Circular economy partners for soluble protein from eggshell membrane material.

streams as inputs to their processes. To that end, they used the Ladder of Circularity of Cramer. The results of using that ladder are shown in Figure 6.2 and Table 6.2.

Then they searched for companies or types of industries that could take the product and the by-product streams as input for their processes. They found potential partners for each by-product and then calculated the Economic Potential of using these by-products by the partners. The results are shown in Figure 6.3. It appeared that for each partner a positive economic potential seems to be feasible.

Bibliography

[1] Ellen MacArthur Foundation. Circular economy scheme. Sourced 24-2-2025. https://www.ellenmacarthurfoundation.org/.

[2] Harmsen J, de Haan AB, Swinkels PL. Product and process design: Driving sustainable innovation. In: Product and Process Design 2024 May 20. De Gruyter.

[3] Cramer J. Circular cities webinar. Resourced from: http://www.nzwc.ca/events/circular-cities-amsterdam/Documents/Webinar-JacquelineCramer-Mar14-17.pdf.

[4] Strohbehn RE, Etzel LR, Figgins J, inventors; Biova LLC, assignee. Solubilized protein composition obtained from avian eggshell membrane. United States patent US 8,173,174. 2012 May 8.

[5] Alves ACO, Polverini S, Arslantürkoğlu E, Consonni F, Tamoor Mughal M. Sustainable integral process design for soluble eggshell membrane product. In: Abstract and Presentation at Process Development Symposium Europe, AIChE. October 21–22. Frankfurt, Germany. 2025.

7 Industrial symbiosis design

7.1 Basics industrial symbiosis

In a systematic review, Yeo defines Industrial Symbiosis as: "Industrial symbiosis (IS) employs a cross-organizational perspective to seek synergistic pairings of one company's waste output to another company's input, enabled by interfirm cooperation through resource and information sharing" [1]. So, it is about using waste of one company as an input from another company. IS, however, is not only cooperating between companies; it can also be cooperation between municipalities and companies In the USA, it is called industrial by-product synergy [2].

Lange [3] describes design interventions in industrial symbiosis but mentions no specific design methods. But it provides no specific design methods. He also focuses on finding the organizational connections between firms and their processes. These organizational aspects of finding these synergistic pairings are outside the scope of this book.

However, there are many successful industrial symbiosis cases described in literature. The representations are often in block-flow diagrams and the descriptions of the connecting streams (flows) between the companies have common characteristics.

Mangan describes an IS where steel slag, a residual waste of steelmaking, is used for making Portland cement. This use was found after the discovery that steel slag contains dicalcium silicate (calcined lime), which constitutes a building block of Portland cement. From this, it can be derived that detailed descriptions of the waste stream compositions

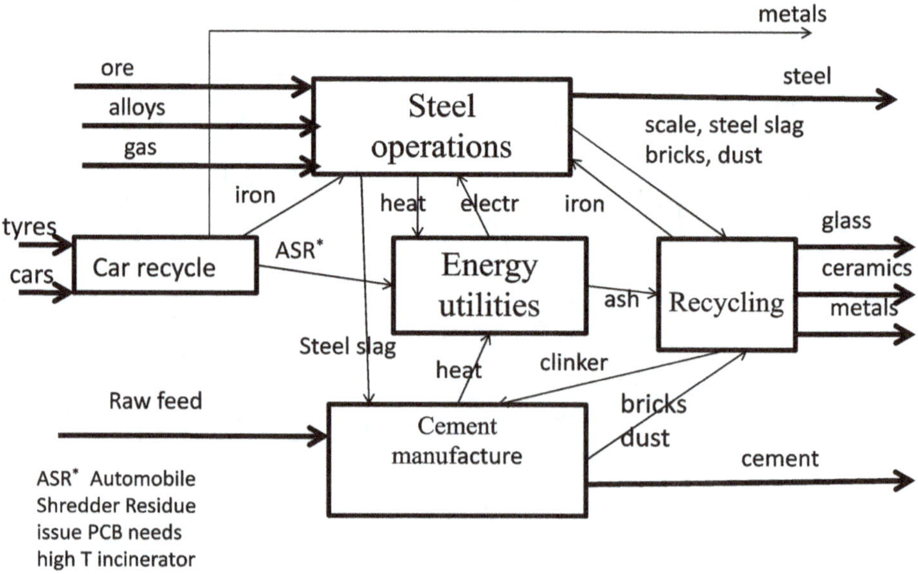

Figure 7.1: By-product synergy Chaparral Steel USA [2].

https://doi.org/10.1515/9783111203256-010

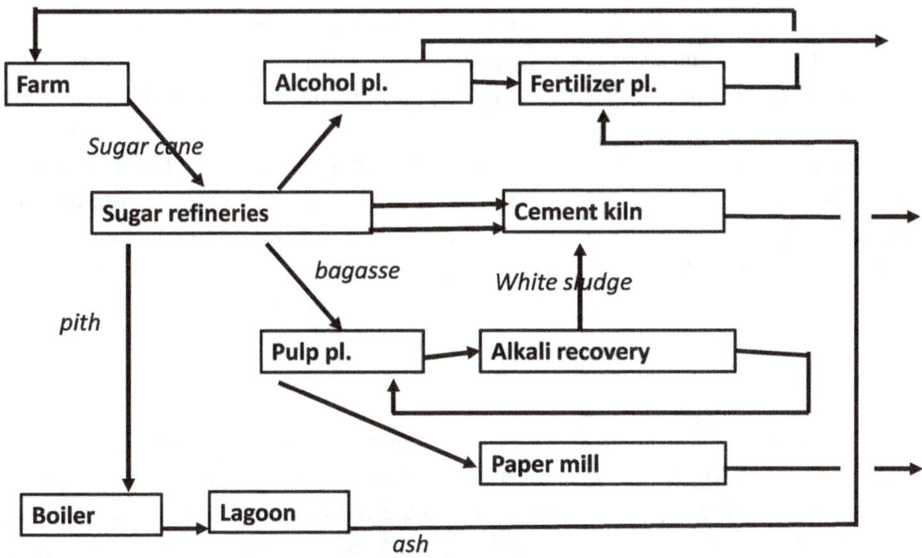

Figure 7.2: Industrial symbiosis Guitang China [5].

are a key element to arrive at a successful synergistic flow paring between companies. The figure shows the block-flow diagram of all partners and streams involved in Figure 7.1 as Zhu describes an industrial symbiosis system where bioprocesses, food processes, a fertilizer, and cement factory are involved [5], as shown in Figure 7.2.

From foregoing cases, it follows that block-flow descriptions are a key communication element of successful industrial symbiosis.

7.2 Industrial symbiosis design method

Industrial symbiosis projects evolve over time after lengthy discussions between partners about options and conditions for cooperation. Neves [7] provides a comprehensive review of industrial symbiosis research. So, a step-by-step design procedure cannot be provided. However, it is clear that block-flow diagrams of companies and their streams are an important method in communication between potential partners. In this way, options for connecting companies are identified.

For new industrial symbiosis projects, these block-flow diagrams have to be generated before the actual symbiosis is implemented. This generation can be done in three steps.

The first step is to represent all existing input and output streams of all companies on the industrial site. Also, nearby municipalities and their streams should be included.

The second step is using these stream descriptions to look for synergies between companies and municipal output and input streams.

The third step is to connect in new ways by block-flow diagrams the companies and municipalities. Each bloc represents the collection of processes belonging to a company. The flows represent all material and energy inputs and outputs flowing to and from the blocks.

These three steps are to be executed by many if not all parties involved. It is clear from a description by Mangan that this execution is not a simple linear process but takes time, as that trust between parties involved has to be built and also stream characterization and stream modifications take time [4].

7.2.1 Company block-flow descriptions

It is essential that each company block describes the collective of all processes belonging to that company. Large industrial sites often have processes, such as wastewater treatment, which are owned by all companies involved. Such a process should also be treated as a separate company identity. This way it aids discussions between companies about which waste streams are involved for potential cooperation.

7.2.2 Output flow descriptions

The output flow descriptions of each company block should first of all contain all mass waste streams. The output mass waste streams can be:
– Gas streams to the atmosphere from furnaces and off-gas treatment process steps. It can be wastewater's waste streams
– Liquid waste streams
– Solid waste streams

Then waste energy streams should be reported. These can be steam streams, or cooling water, or waste streams that can be combusted to generate electricity and/or heat, or an output stream at higher than ambient temperature, such as a water output stream from a wastewater plant, which is 12 °C higher than ambient temperature and can then be used as input to district heating [6].

7.2.3 All input flow descriptions and alternative inputs

The input flow descriptions of each company should all be made explicit. It is also useful to define these streams not only in their composition, but also what function they have in the processes. For cement production, for instance, an input stream function can be alkalinity. This functional description can generate alternative (waste) input streams to be considered, such as what happened in the Chapparal Steel industrial symbiosis where waste alkaline steel slag was used for cement production, when the cement producer mentioned that he needed alkalinity [4].

7.3 Example case: Industrial symbiosis of the city Terneuzen and Dow Chemical Company

The Dow Chemical Company has a large chemical production complex in Zeeuws-Vlaanderen of the Netherlands. In the mid-1990s, Dow enlarged it production capacity significantly. This expansion required expansion of the site water facilities. The region Zeeuws-Vlaanderen is structurally short of fresh water. Dow Chemical headquarters already had an industrial symbiosis policy and a managerial method implemented. Therefore, Dow, Evides; the industrial water supplier, the Zeeuws-Vlaanderen Water Board, and the community Terneuzen joined their efforts to sustainably control the regional water balance.

Several studies were performed resulting in a plan to convert municipal wastewater by a wastewater treatment plant and transport the resulting water by a new pipeline of 6 km to the Dow site. By membrane filtration, the water was desalted to produce boiler feed water. The plan was implemented successfully. Figure 7.3 shows the block-flow diagram of the parties involved and their resulting streams.

This whole industrial symbiosis, involving 2.5 million m^3/year wastewater to boiler feed water conversion saved 65 % of energy compared to using brackish surface water from the Westerschelde river. It also reduced the waste salt stream of the reverse osmosis membrane plant when river water was used as water source.

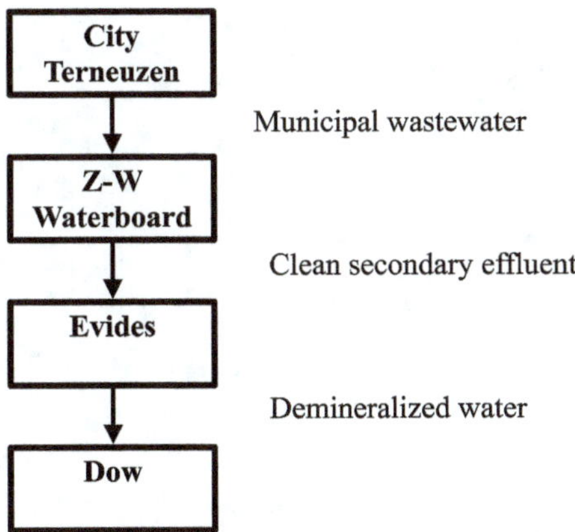

Figure 7.3: Industrial symbiosis Dow and municipality Terneuzen.

This industrial symbiosis project received in 2007 the European Responsible Care Award. The jury mentioned that: "This innovative water project demonstrated a commitment to the local community and to the environment and showed the industry is going above and beyond what is required" [2].

There are many regions where water is produced from sea water and where municipal wastewater is also available. So, this type of industrial symbiosis can be applied elsewhere as well.

Bibliography

[1] Yeo Z, Masi D, Low JS, Ng YT, Tan PS, Barnes S. Tools for promoting industrial symbiosis: A systematic review. Journal of Industrial Ecology. 2019 Oct;23(5):1087–108.
[2] Wu Q. Industrial ecosystem principles in industrial symbiosis: By-product synergy. In: Harmsen J, Powel J, editors. Sustainable Development in the Process Industries, Cases and Impact. Hoboken: J. Wiley; 2010.
[3] Lange KP, Korevaar G, Oskam IF, Herder PM. Developing and understanding design interventions in relation to industrial symbiosis dynamics. Sustainability. 2017 May 16;9(5):826.
[4] Mangan A, Olivetti E. By-product synergy networks: Driving innovation through waste reduction and carbon mitigation. In: Harmsen J, Powel J, editors. Sustainable Development in the Process Industries, Cases and Impact. Hoboken: J. Wiley; 2010.
[5] Zhu Q, Lowe EA, Wei YA, Barnes D. Industrial symbiosis in China: A case study of the Guitang Group. Journal of Industrial Ecology. 2007 Jan;11(1):31–42.
[6] Gruber-Glatzl W, Brunner C, Meitz S, Schnitzer H. From the wastewater treatments plant to the turnstiles of urban water and district heat networks. Frontiers in Sustainable Cities. 2020;2:60.
[7] Neves A, Godina R, Azevedo SG, Matias JC. A comprehensive review of industrial symbiosis. Journal of Cleaner Production. 2020 Feb 20;247:119113.

Part III: **Chemical reaction engineering**

8 Chemical reaction engineering single-phase

8.1 Basic chemical reaction engineering single-phase

8.1.1 Main elements of chemical reaction engineering

The chemical reaction engineering method consists of determining reaction chemistry, reaction kinetics, and then use that information first of all to explore the best reactor concept for highest product yield on feed components, and second, to explore the lowest investment cost for the same high yield. The background reasoning is that for most processes the feedstock costs are the largest element in the total product cost. Hence, the highest yield of product on feedstock is the most important economic design criterion.

8.1.2 Chemistry and chemical stoichiometric equation

All chemical reaction engineering design activities need reaction chemistry and reaction rate kinetics as input information. This information is obtainable from experiments with lab-scale reactors. The reactors need to have a uniform temperature and concentrations in the reaction space. Their operation can be batch or continuous.

By analyzing the lab-scale experiments on the disappearance and formation of chemical components, the reactions involved can be determined and the stoichiometric equation for each reaction can be determined.

The general chemical stoichiometric equation for a chemical reaction is

$$aA + bB = pP + qQ \tag{8.1}$$

The convention is that on the left-hand side of the equation the reactants are stated, and the products are on the right-hand side, with the stoichiometric coefficients a, b, p, q.

It is an atom balance. As in chemical reactions, atoms are not changed; the number of each atom type stays the same. This means that the number of each atom on both sides of the equation is the same. The stoichiometric coefficients, ab, p, q, are to be chosen such that this is then indeed the case.

An example is the combustion of methane in which methane reacts with oxygen, forming carbon dioxide and water. The chemical stoichiometric equation is then given by equation (8.2):

$$CH_4 + 2\,O_2 = CO_2 + 2\,H_2O \tag{8.2}$$

For C, H, and O, the number of atoms is now indeed the same on both sides of the equation.

Often several reactions take place. For each of these reactions, the chemical stoichiometric equation then has to be determined.

https://doi.org/10.1515/9783111203256-012

With these equations, a chemical reaction network can be determined. The network may contain parallel reactions of reactants to the products and to by-products and consecutive reactions of the product to other by-products [1].

In some cases, many reactants are present in the feed and many product components are formed such as in biomass pyrolysis. In those cases, the reactants can be lumped into cellulose, hemicellulose, and lignin, for which atom compositions can be determined. The product components can also be lumped into desired product components and undesired components such as coke. Also, for these lumped components atom compositions can be determined by chemical analysis. A preliminary chemical reaction network of consecutive and parallel reactions can be postulated and be validated by experiments.

8.1.3 The chemical reaction rate

The chemical reaction rate r is defined as the rate at which species is formed or disappears by a chemical reaction, per unit reaction volume, with S. I. units $mol/(s.m^3)$. The reaction rate r is an intrinsic variable, similar to temperature. So, it is a local variable inside a reactor, with in general different values depending on the local temperature and local chemical species concentrations.

Each reaction rate furthermore belongs to a chemical (stoichiometric) equation.

For the generic chemical stoichiometric equation: $aA + bB = pP + qQ$.

The reaction rates of the species A, B, C, D for this are then linked to each other by [2]

$$-r_A/a = -r_B/b = r_C/c = r_D/d \tag{8.3}$$

So, the reaction rate can be determined for one of the four reaction rates. The others directly follow from the stoichiometric coefficients using expression (8.3).

The most reliable laboratory set up for determining chemical reaction rate is a continuously operated fully back-mixed reactor. In such a gradientless reactor, the reaction rate r is directly obtained from the steady-state mole balance over the input and output and the reactor reaction fluid volume V. The reaction rate is determined by expression (8.4)

$$-r = (F_{in} - F_{out})/V \tag{8.4}$$

in which F_{in} is the molar flow in (mol/s) and F_{out} is the molar flow out of the reacting component. The negative sign in front of r is here for a case where the component reacts away. So, r has a negative value if $(F_{in} - F_{out})$ has positive value, meaning that the component reacts away.

By performing many experiments for different concentrations of the reactants and product species, the reaction rate can be correlated with these concentrations—kinetic expressions.

Typical kinetic expressions for the chemical equation given above in this section are the following (see Table 8.1).

Table 8.1: Chemical reaction rate expressions for single phase kinetics.

Rate expression		Rate expression name
$r = kA$	(8.5)	First order in A
$r = kA^2$	(8.6)	Second order in A
$r = kAB$	(8.7)	First order in A and B

The reaction rate in general also depends on temperature. Often the temperature dependance can be expressed with an Arrhenius-type expression.

The temperature dependency of the reaction rate is put into the reaction parameter k. It is often given by the Arrhenius law expression (8.8):

$$k = k_0 e^{-E_a/RT} \tag{8.8}$$

where:

k_0 is the preexponential constant, in units depending on the kinetics,

E_a is the activation energy, in J mol^{-1},

T is the temperature of the reaction control volume in Kelvin,

R is the universal gas constant with its value and dimensions 8.314 J mol^{-1} K^{-1} [3].

The reader is, however, warned that different dimensions are still used as well. It is particularly important that the activation energy E_a applied has the same S. I. dimensions as R.

However, it is better to use a modified version of the Arrhenius expression (8.9)

$$\text{Log}(k/k_{ref}) = (E_a/R)(1/T_{ref} - 1/T) \tag{8.9}$$

In this expression, k_{ref} is the reaction rate at the reference temperature T_{ref}. This reference temperature is then chosen in the experimental range. In this way, the two parameters k_{ref} and E_a, fitted to the experiments, will be independent from each other. In the original Arrhenius expression, k_0 is defined at an infinitely high temperature, so far beyond the experimental range. This means that k_0 cannot be accurately determined and in practice its value correlates to the value of E_a, resulting in a less reliable Arrhenius expression.

8.1.4 Plug-flow and back-mixed reactor concepts

For single phase reactors, with no catalyst or a homogeneous catalyst, the most important characteristic for the reactor performance is its residence time distribution (RTD). We treat here two reactor continuously operated ideal concepts, the plug flow reactor, and the back-mixed reactor. These two ideal concepts are very different in RTD and thereby for most chemical reaction schemes very different in performance. We will first describe the residence time distributions and resulting concentration profiles inside these reactor concepts and then describe for a few reaction chemistry schemes the performance differences.

For the plug flow reactor, all fluid elements entering the reactor will have the same residence time. This means that for each fluid element the reaction performance is the same. The concentration of reactants will drop with increasing residence time, and the concentration of product components will increase with residence time.

For the back-mixed reactor, all fluid elements entering the reactor instantaneously mix with all fluid elements already in the reactor, resulting in a uniform concentration of reactants and products inside the reactor. The reactor outlet concentrations will therefore also be the same as the concentrations inside the reactor.

Figure 8.1 shows the concentration profiles of reactant A and product P for the plug flow reactor and the back mixed reactor.

For the plug flow reactor, the concentration of A drops along the length of the reactor, and the product concentration increases along the length of the reactor.

For the back mixed reactor, the concentration A is uniform along the length of the reactor and the concentration of P is uniform along the length of the reactor.

These differences in concentration profile have consequences for the reactions taking place in these reactors.

If the reaction rate depends on the concentration of reactant A, then the plug flow reactor will show a deeper conversion than the back mixed reactor for equal reactor volume. If the back mixed reactor is designed for the same conversion as the plug flow reactor, then a much larger volume will be needed. If, for instance, the conversion of A for a first-order reaction in A in the plug flow reactor is 99 %, then the volume of the back mixed reactor has to be a factor 15 larger to reach the same conversion.

If for this case the product P degrades to undesired by-products, then the back mixed reactor will show far more by-product formation, for two reasons. The first reason is that the residence time in the back mixed reactor is a factor 15 larger (for the same conversion); hence, even for a zero-order consecutive reaction the by-product formation will be a factor 15 larger. If the consecutive reaction increases with increasing concentration of P, then the by-product formation in the back mixed reactor will be even larger than the factor 15.

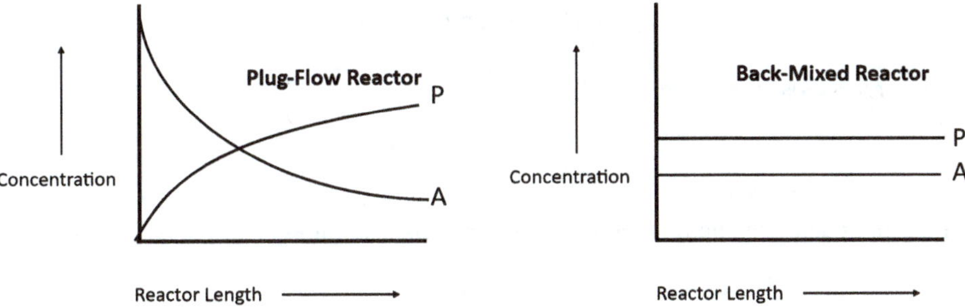

Figure 8.1: Concentration profiles reactant A and product P of a plug flow reactor and a black-mixed reactor.

8.2 Reactor concept design method

8.2.1 Reactor concept choices for conversions and degree of back-mixing

Guidelines for choosing the best reactor concepts for degrees of back-mixing (residence time distribution) are given below for some kinetic cases.

First of all, the reaction kinetics have to be obtained. These can be obtained from literature sources or have to be determined experimentally. Here, we treat some reaction kinetics rate expressions and provide guidelines for reactor concept design for each of these kinetics.

Reactor concept case: Single irreversible reaction

$$\text{chemical equation:} \quad aA = pP\ (+qQ) \tag{8.10}$$

$$-r_A = kA^n, \quad n \geq 0 \tag{8.11}$$

Design guidelines

Design guideline 1. Choose such a deep conversion of A that it is below the specification of A in the product stream P and so that therefore separation of A from P is not needed.

Design guideline 2. Choose a plug flow reactor. Only then is a very deep conversion obtainable in a limited reactor volume. Theoretically, this guideline holds for $n > 0$, but even if $n = 0$ then a plug flow reactor is preferred for very low concentrations of A. Even a zero-order reaction nearly always changes to higher-order reaction in A. Hence, a plug flow reactor concept is needed.

A very useful correlation for determining a plug-flow reactor-concept design is the correlation derived by Mears [3]. The correlation links the desired conversion X to the minimum value of the Peclet number required for plug flow behavior and maximum allowed error in conversion. The correlation (8.12) is given below.

$$Pe_{min} = (100/error)\, n \ln(1/(1-X)) \qquad (8.12)$$

For an error = 5 %, the correlation becomes then expression (8.13)

$$Pe_{min} = 20n \ln(1/(1-X)) \qquad (8.13)$$

For a pipe reactor, the actual Peclet number can be is obtained from Westerterp [4]

$$Pe = (L/D)\, Bo \qquad (8.14)$$

in which

$$L = \text{the reactor length} \quad \text{and} \quad D = \text{the reactor diameter}$$

The Bodenstein number Bo is defined by (8.14) as

$$Bo = v.D/Dax \qquad (8.15)$$

The Bodenstein number depends on the Reynolds number given by equation (8.15).

$$Re = \rho v D/\eta \qquad (8.16)$$

For Re > 10,000; Bo = 2.
This results then in

$$Pe = 2(L/D) \qquad (8.17)$$

The theoretical minimum number of back-mixed reactors in series N_{minmix} can be approximated from an expression by Elgett [5].

$$N_{mix} = (Pe/2) + 1 \qquad (8.18)$$

Replacing Pe_{min} by $N_{mix\ min}$ in the expression for Pe_{min} results in

$$N_{minmix} = 0.5 + 10 \ln((1/(1 = X)) \qquad (8.19)$$

In general, the calculated value will not be an integer. The minimum number of back mixed reactors in series should be obtained by rounding of the calculated figure.

Reactor concept case: Single irreversible reaction with chemical equation aA + bB = pP (+qQ)

$$-r_A = kA^n\ B^m \qquad (8.20)$$

Design guideline 3. Choose for which component A or B a deep conversion is desired to avoid separation from product P. Then choose this component to be the limiting reactant and choose the other component to feed to the reactor in excess. This choice ensures a high reaction rate.

Design guideline 4. Choose a very deep conversion of the limiting reactant. This choice avoids a separation step.

Design guideline 5. Choose a plug flow reactor to obtain the deep conversion in a limited reactor volume.

Reactor concept case: Consecutive reactions with chemical equations A = P and P = X
P is the desired product, and X is the undesired by-product.

Design guideline 6. Choose a single pass limited conversion of A such that X is hardly formed. Separate A from product P and recycle A to the reactor inlet.

Design guideline 7. Choose a plug flow reactor. This reduces the reaction of P to X, compared to a back-mixed reactor.

Reactor concept case: Parallel reactions (1) A = P and (2) A = X

$$-r_{1A} = k_1 A^n \tag{8.21}$$

$$-r_{2A} = k_2 A^m \tag{8.22}$$

For kinetics expressions given by equations (8.19) and (8.20), it is useful to take the ratio r_{1A}/r_{2A} and from that ratio it is clear how to obtain a high value of the ratio. With these ratio guidelines, 8, 9, and 10 are directly derived.

Design guideline 8. If $n > m$, keep concentration A high; hence, go for a plug flow reactor type with a limited conversion of A, separate it from P and then recycle A.

Design guideline 9. If $n < m$, keep concentration of A low; hence, go for a back mixed reactor type and go for deep conversion of A [6].

Design guideline 10. If $n < m$, then in addition to guideline 9, separate product P from A and recycle part of it to the reactor to further reduce the concentration of A [6].

8.2.2 Example reactor concept design for autocatalytic kinetics

The example case treated here is an autocatalytic reaction, in which the product P catalysis the reactor of A to P. This example is chosen to reveal that the choice for a reactor concept is not between a plug flow reactor and a back mixed reactor, but that a combination in the optimum sequence is the best choice.

Reactor concept case: Autocatalytic reaction by product enhancing reactant reaction

The chemical equation is aA = pP.

The autocatalytic kinetic reaction rate expression is typically given by equation (8.23):

$$-r_A = kAP \tag{8.23}$$

This kinetic rate expression shows that at higher P concentration the reaction is faster. For this reason, the reaction is called autocatalytic. The reaction is "catalyzed" by its own product. The reaction rate expression also implies that if P = 0, then the reaction rate is zero, so the reaction does not start.

Design guideline 11. For autocatalytic reactions, choose a back-mixed reactor up to a certain conversion.

Because a high P concentration everywhere in the reactor results in a high reaction rate, it means that a back mixed reactor is favored up to a certain conversion of A.

Design guideline 12. For autocatalytic reactions, choose a plug-flow reactor after the back mixed reactor.

At a high conversion of A, the concentration of A becomes low, so the reaction rate then drops. The best reactor concept design is therefore a back-mixed reactor followed by a plug flow reactor. The latter reactor then facilitates a deep conversion of A, while having a high concentration of P.

The back mixed reactor concept at the front will need less volume for a certain conversion than a plug flow reactor. The plug flow reactor would furthermore need some recycling of product P to the inlet to start the reaction.

The back mixed reactor needs some amount of product at the start-up of the reactor but when steady-state operation is achieved, it does not need any feed of P anymore.

The optimum conversion for the back mixed reactor can be found by plotting $1/-r_A$ versus the conversion of A; expressed as X_A. The curve of $1/ = r_A$ versus X_A drops until a minimum value is reached and then it increases with higher X_A. The value of X_A at minimum point is the optimum point for the back mixed reactor conversion [2, p. 141]. For stoichiometric coefficients, a = p = 1, the optimum conversion for the back mixed reactor is $X_A = 0.5$.

The plug flow reactor should then be designed for such a deep conversion of A that separation of A from P is not needed. For that design, the reactor system of the back mix followed by a plug flow reactor is not only the best reactor concept design but also the best process concept design.

There are several reaction types that show autocatalytic behavior for which this example case is relevant. These are fermentations where the microorganism cell growth kinetic rate is proportional to the microorganism cell concentration and to exothermic adiabatic reactions such as in furnaces where the reaction rate increases with conversion, due to the higher temperature. For both of these systems, a back mixed reactor followed by a plug flow reactor is the best solution for the smallest overall reactor volume required and for achieving deep conversion of the feed components [2].

8.2.3 Reactor concept design for temperature choice

The choice of the reaction temperature is also important in concept design. That choice also influences the reaction selectivity.

For parallel reactions between the desired product and the undesired by-products, the reaction rate ratios have to be examined. If that ratio is high for high temperatures, then of course a high temperature should be chosen. Using the Arrhenius expression for temperature effects on the reaction, this means that the activation energy for the main reaction is higher than the parallel side reaction. If it is the other way around, then a low temperature should be chosen.

For reactions where the by-product is formed by a consecutive reaction and the activation energy for the main reaction is higher than that for the consecutive reaction, then the reaction temperature should be high and as a compromise between reaction rate a by-product formation an increasing reaction profile could be chosen. If the activation energy of the desired reaction is low, then a low temperature and as compromise a decreasing temperature profile can be chosen.

How to design for reactors with heat effects is treated in Chapter 14.

For most designs, the highest product yield on feedstock is desired. This means that the reaction temperature should be chosen such that by-product formation is minimized.

Guidelines below summarize these reasonings.

Design guideline 13. Choose high reaction temperature if E_a main > E_a by-product.

If the activation energy of the main reaction is higher than that of the by-product reaction then the reaction temperature should be high, so that the main reaction is favored over the by-product reaction

Design guideline 14. Choose a low reaction temperature if E_a main < E_a by-product.

8.3 Example case: Industrial ethylene glycol reactor design

Melhem provides all kinetic data for the main reaction of ethylene oxide water to ethylene glycol. He also provides kinetics for the consecutive reaction of ethylene glycol to di-ethylene glycol and higher glycols [8].

The reactions and kinetics are

Main reaction:

$$\text{Ethylene Oxide} + H_2O = \text{Ethylene Glycol (EG)} \quad \Delta H = -389 \text{ KJ/mol}$$

The kinetics of the reaction are first order in EO and water.

$$r_1 = -k_1[EO][H_2O] \tag{8.24}$$
$$k_1 = k_{10}e^{-E_a1/RT} \tag{8.25}$$
$$E_a/R = 9{,}525 \text{ K} \tag{8.26}$$

By-product reactions

$$EG + EO = \text{Di-ethylene Glycol (DEG)}$$
$$DEG + EO = \text{Tri-ethylene Glycol (TEG)}$$

The kinetics of these reactions are first order in EO and DEG and TEG. The reactive intermediate complex from EO and (alcohols water inclusive) for the main reaction and the by-product reactions is the same. This means that the activation energy for all reactions is the same [8].

Reactor feed ratio design

To minimize the undesired by-product reactions over the main reaction water is to be provided in surplus, so that the main reaction is much faster than the by-product reactions.

The surplus water, however, has to be separated from the product. This is executed by distillation. So, there is an economic optimum for the water to EO feed ratio, which depends on the economic value of the by-products DEG and TEG. An optimum ratio water/EO is not reported, but a Water/EO molar ratio = 20 (mol/mol) means that around 5 % of DEG is formed, which could be starting point for a detailed economic optimization of separation investment and energy cost versus the difference in economic sales potential difference between EG and DEG.

The feed flows to the reactor can now be determined.

The design case capacity is 750 kton/y (Shell Singapore MEG capacity).

Assumed is running time 8000 h/per year = 28.8 s.

This means that the MEG production rate is 750 / 28.8 = 26 kg/s.

The EO feed = 35/(18 + 35) = 0.66 × 26 = 17 kg/s.

The water is 20 mol H_2O/mol EO = 20 × (18/35) × 17 = 175 kg/s.

The volumetric flow is then 0.18 m^3/s.

Reactor RTD design

It is desired that no EO is present in the outlet of the reactor so that no separation of EO is needed. Moreover, EO is very toxic [7], so keeping process volumes with EO present small is highly desired. The toxicity limit in air = 0.1 ppm. So, this means that a very deep single pass conversion of EO in the reactor is desired. We take this as limit for the reactor concentration exit. The design conversion of EO is then set at X = 0.9999999 or $(1 - X) = 0.0000001$.

Water is present in large surplus, so the outlet of the reactor is mainly water, and so this conversion limit for safety reasons is very conservative. A factor 10 lower would also still be okay.

The deep conversion of EO is best achieved in a pipe flow reactor at turbulent conditions with Renolds number > 10,000.

Using expression (8.12) results in a minimum Peclet number value Pemin = 320.

For turbulent pipe flow Pe = 2 L/D. This means that the length/pipe diameter ratio should be L/D > 160.

From the required residence time for the main reaction in the plug flow reactor. the reactor volume is determined. With the L/D > 160, a design space for the concept design can be determined. In the detailed engineering stage. the reactor design can then be further optimized within this design space.

Reactor temperature design

As the main reaction and the by-product reactions have the same activation energy. the reaction temperature can be freely chosen. An economic optimum can be determined in the development stage, taking into view that a high temperature means a small reactor volume but a high pressure to keep water in the liquid phase.

Bibliography

[1] Schaschke C. A Dictionary of chemical engineering. Oxford: OUP; 2014.
[2] Levenspiel O. Chemical Reaction Engineering. Hobokn: John Wiley & Sons; 1998.
[3] Mears DE. The role of axial dispersion in trickle-flow laboratory reactors. Chemical Engineering Science. 1971;26:1361.
[4] Westerterp KR, van Swaaij WP, Beenackers AA, Kramers H. Chemical reactor design and operation. Chicester: John Wiley and Sons; 1984.
[5] Elgeti K. A new equation for correlating a pipe flow reactor with a cascade of mixed reactors. Chemical Engineering Science. 1996 Dec 1;51(23):5077–80.
[6] Douglas JM. Conceptual Design of Chemical Processes. New York: McGraw Hill; 1988.
[7] The National Institute of Occupational Health and Safety. Sourced 5 March 2025. https://www.cdc.gov/niosh/idlh/75218.html#:~:text=Immediately%20Dangerous%20to%20Life%20or%20Health%20Concentrations%20(IDLH)&text=NIOSH%20REL%3A%20%3C0.1%20ppm%20(,policy%20%5B29%20CFR%201990%5D.
[8] Melhem GA, Gianetto A, Levin ME, Fisher HG, Chippett S, Singh SK, Chipman PI. Kinetics of the reactions of ethylene oxide with water and ethylene glycols. Process Safety Progress. 2001 Dec;20(4):231–46.

9 Gas-liquid reactors

9.1 Basics gas-liquid reactor type selection

9.1.1 Overview of reactor type selection

In gas-liquid reactors, one or more of the reacting components are transferred from the gas phase to the liquid phase, where they react with themselves or with other components already in the liquid phase.

The transfer of the gaseous component(s) can be a rate limiting factor. If that is the case, then the reactor size is determined by the mass transfer performance of the reactor type chosen. If mass transfer is such a limiting factor that the reaction only takes place in the boundary layer at the gas-liquid interface and a consecutive reaction to an undesired side product takes place, then the product yield on feedstock is also negatively affected.

If mass transfer is the limiting factor, then reliable mass transfer coefficients for the commercial scale are needed. However, mass transfer coefficients and their correlations have in general determined by water-air experiments at small scale; hence, these correlations are often not reliable. For this reason, and because this book has a limited scope, the design procedure of this chapter is based on selecting a reactor type, for which gas-liquid mass transfer is not the limiting factor.

Residence time distributions of the gas phase flow and the liquid phase flow can also affect reaction conversion and product formation selectivity. These performance phenomena are also scale dependent. So, reactor type selection should also include the desired residence time distribution. For many reactor types, residence time distribution information is available and conservative estimates of their commercial scale residence time distribution behavior can be made.

Sometimes the reactor needs to have heat exchange to keep the reactor temperature at the desired value. So, then also reactor type selection involves reactor type heat exchange capability.

Section 9.1.2 describes the generic theory of gas-liquid mass transfer and reaction. With that information, a simple procedure is presented to determine the minimum requirement for mass transfer to obtain the condition of no mass transfer limitation.

Section 9.1.3 describes most common industrially applied types with their mass transfer and residence time distribution characteristics and their capability for heat exchange.

Section 9.2 describes a reaction selection procedure for a reactor type for the following:

- No mass transfer limitation
- Optimal residence time distribution
- Heat exchange

https://doi.org/10.1515/9783111203256-013

Section 9.3 describes an industrial case of reactor type selection in view of mass transfer, residence time distribution, and heat exchange.

9.1.2 Gas-liquid mass transfer and reaction theory

The design approach for gas-liquid presented here is to first describe the mass transfer from the gas bubbles to the liquid phase, and then derive a generic criterion for designs without mass transfer limitations. For the mass transfer description, we take a slab anywhere in reactor consider the mass transfer from gas phase (bubbles) to the liquid phase. The concentration profile in the gas and liquid phase is shown in Figure 9.1 for a case with significant mass transfer limitations, just to illustrate the subject at hand.

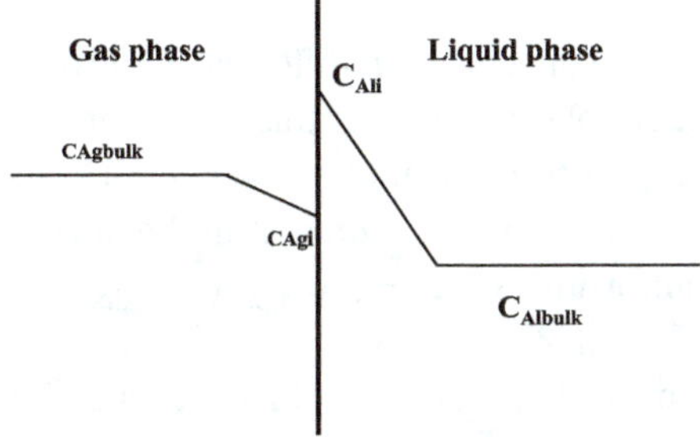

Figure 9.1: Concentration profiles in a gas-liquid-phase reactor mass transfer from gas phase to liquid phase.

The mass transfer rate from the gas phase to the liquid phase (where the reaction takes place) has two steps. The first step is mass transfer from the gas bulk to the gas-liquid interface via a boundary layer near the interface, with a mass transfer coefficient k_g. At the interface, the component A on the gas side is in equilibrium with the liquid side. The second step is mass transfer from the liquid side interface to the bulk of the liquid through the liquid boundary layer, with a mass transfer coefficient k_l.

The overall flux of A from the bulk of the gas phase to the bulk of the liquid phase J_A is then given by Westerterp [1]. By multiplying the flux per square meter interface by the specific surface area, a mass transfer rate per unit liquid volume is obtained.

$$aJ_A = a\{k_l/(1 + (mk_l/k_g)\}(mC_{Agbulk} - C_{Albulk}) \tag{9.1}$$

For clarity in the further reasoning, a Mass Transfer Performance (MTP) is introduced:

$$MTP = a\{k_l/(1 + (mk_l/k_g)\} \tag{9.2}$$

This represents the mass transfer performance characteristic of the reactor.

Also, a concentration driving force for mass transfer ΔC is introduced:

$$\Delta C = (mC_{Agbulk} - C_{Albulk}) \tag{9.3}$$

With this the mass transfer expression (9.1) is now rewritten as

$$aJ_A = MTP. \Delta C \tag{9.4}$$

At steady state, the molar flow of component A ($J_A a$) from the gas phase to the liquid phase is equal to the reaction rate $- r_A$. (This is the case because the molar flow and reaction rate are expressed per unit liquid volume.)

So,

$$J_A = MPT. \Delta C = -r_A \tag{9.5}$$

Now mass transfer limitations are absent when the concentration differences between the for the mass transfer to happen are small. A simple criterion for this to happen can now be derived. Let us assume that the effect of mass transfer limiting the overall reaction rate should be less than 5 %. This means that the concentration of A in the bulk of the liquid should be maximally 5 %, then at equilibrium. So, $\Delta C = (mC_{Agbulk} - C_{Albulk}) = 0.05C_{Albulk}$.

Substituting the driving force ΔC into the expression J_A, a result in an expression for no mass transfer limitation.

$$MTP_{\text{no mass transfer limitation}} = (1/(0.05C_{Albulk})) - r_A = 20r_A/C_A \tag{9.6}$$

The concentration of A to be used in the reaction rate expression is the bulk concentration at equilibrium:

$$C_{Ali} = mC_{gbulk} \tag{9.7}$$

For a first-order reaction in A: $-r_A = k_1 C_A$, it follows then that

the minimum mass transfer parameter value for no mass transfer = $(1/0.05) = 20k_1$.

If the mass transfer of the liquid side is determining the overall mass transfer, then the rule of thumb can be used that $k_l a = 20k_1$ is the minimum mass transfer parameter value needed to avoid mass transfer limitation.

It is important to note that the MTP contains three separate parameters: a gas phase mass transfer k_g parameter, a liquid phase parameter k_l, and a specific surface area parameter a. All three are independent from each other. However, experimental research results often report lumped $k_l a$, values. In some cases, however, for high values of m, gas phase mass transfer also plays a role, but that effect is then hidden in the lumped experimental result. Needless to say, such lumped mass transfer parameter values are unreliable.

Design for no mass transfer limitation in liquid film layer

Some reactions are so fast that avoiding mass transfer limitation by design, as described in Section 9.1.2 is not feasible. Then the reaction takes place in the liquid film layer next to the interface. A reacting component diffusing in the film layer is reacted away to zero concentration. The reacting component from the bulk liquid phase diffusing in the film layer is also reacted away to some extent. The resulting product formed will have a higher concentration in the liquid film then in the liquid bulk as it is also transported by diffusion only.

The dimensionless numbers that govern this reaction and mass transfer are the Hatta number and the f_{max} maximum enhancement factor.

This means first of all that the reaction inside the film layer should be slow compared to the mass transfer rate. The ratio of reaction rate and mass transfer is defined by the Hatta number for the reaction:

$$A + B = P$$

With the kinetic expression,

$$-r_A = kC_A^n C_B^q$$
$$Ha^2 = (2/(n+1)kC_{Ali}^{n-1}C_{Bbulk}^q D_A)/k_l^2 \tag{9.8}$$

Figure 9.2 shows three conditions depending on the Hatta number value.

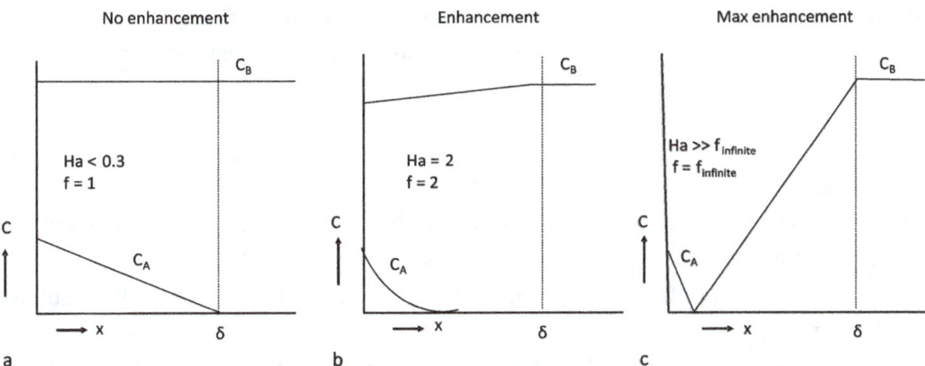

Figure 9.2: Mass transfer with reaction for G-L reactors.

Figure 9.2(a) show for Ha < 0.3 shows that the concentration of component A drops in the film layer in a straight line towards the bulk of the liquid, while the concentration of component B in the film layer is the same as the bulk concentration. This is the regime of no mass transfer enhancement. For this condition, the theory of mass transfer plus reaction of the previous section can be applied.

Figure 9.2(b) for Ha = 2 shows that the concentration of component is curved. The slope of the curve at the gas-liquid interface is steeper than for Figure 9.2(a). This means the mass transfer rate of component a is enhanced by the reaction. This is expressed by the enhancement factor f. The value of f = 2; the same as the Hatta number. This regime is called enhanced mass transfer. The mass transfer coefficient k_l is enhanced by the value of f.

Figure 9.2(c) for Ha $\gg$ f_∞ shows that the concentration of B drops to zero in the film layer. This means that the mass transfer of component B is the rate limiting factor for the reaction plus mass transfer in the film layer. The enhancement factor f is then equal to the maximum enhancement factor. So,

$$f = f_\infty \tag{9.9}$$

$$f_\infty = 1 + D_B C_{Bbulk}/D_A C_{Ai} \tag{9.10}$$

If the product reacts further to undesired by-products, then the diffusion of the product in the film layer also becomes important. Diffusion coefficients have in general similar values. This means that if diffusion of B in the film layer is limiting, then the diffusion of product away from the film layer to the bulk of the liquid will be limiting. For consecutive reactions of the product, this means that by-product formation will be higher. This is undesired to the mass transfer coefficient k_l should be increased so that this condition is avoided.

The mass transfer coefficient should be designed such that Ha $\ll$ f_{max}. This often means that a special reactor type has to be chosen with high k_l values.

Such reactors are the rotating packed bed and the spinning disk reactor. Details are provided by Stankiewicz [2].

Symbol definitions
- a is the specific interfacial area between bubbles and liquid. It is expressed here as area per unit liquid volume.
- J_A is the molar flux of A from the bulk of the gas phase to the bulk of the liquid phase $(mol/(m^2 \, s))$.
- k_l is the mass transfer coefficient for the transfer from liquid interface to the liquid bulk (1/s).
- k_g is the mass transfer coefficient for the gas phase to the gas side interface (1/s).
- C_{agbulk} is the concentration of A in the bulk of the gas phase.
- $m = C_{Ali}/C_{Agi}$ is the equilibrium coefficient for A at the interface (mol liquid/mol gas).

9.1.3 Industrial gas-liquid reactor types and their performance phenomena

Six gas-liquid reactor types, which are applied at commercial scale are here described with a focus on their performance phenomena, mass transfer, residence time distribution, and heat exchange. These descriptions can then be used to apply the selection procedure of Section 9.2.

9.1.3.1 Bubble column reactor performance phenomena

Bubble column reactors have a low mass transfer rate performance. So, their application is limited to slow reactions. For those cases, the bubble column is an attractive option, as it can be designed for large reactor volumes at low investment cost, as it is a relatively simple vessel with a high liquid hold-up. Liquid flow staging is feasible. Also, heat exchange is also feasible. Here is a simple assessment method for these three performance items.

Gas-liquid mass transfer bubble column

The gas-liquid mass transfer coefficients of bubble columns are low, and also dependent on the liquid surface tension and liquid viscosity, which determine the bubble size and thereby the specific surface area a. The liquid side mass transfer coefficient k_l is dependent on the liquid diffusion coefficient, and the gas side mass transfer k_g depends on the gas diffusion coefficient. The overall mass transfer also depends on the value of the solubility equilibrium parameter m.

Here are two procedures to conservatively determine the mass transfer for a given reaction system.

Theoretical procedure for bubble size

The maximum bubble size in a turbulent field can be determined from the expression by Li [3].

$$d_{Bmax} = 1.7(\sigma/\rho)^{0.6}P_m^{-0.4} \tag{9.11}$$

d_{Bmax}:   Maximum bubble size in a turbulent field
σ: Interface tension between gas and liquid
ρ: Liquid density
P_m: Power dissipation per unit mass

The power dissipation of a bubble column per unit liquid mass P_m is given by

$$P_m = v_{sup}\, g \tag{9.12}$$

Experimental procedure for bubble size

Take a transparent small bubble column and feed it with real liquid and with nitrogen gas. Visually observe or measure the bubble size d_B.

Then determine the specific surface area a for spherical bubbles by

$$a = \varepsilon_B 6/d_B \tag{9.13}$$

in which ε_B is the bubble volume fraction.

The mass transfer coefficient k_l is then conservatively estimated by

$$Sh = 2 \tag{9.14}$$

By filling in the values of d_B and D_l in expression (9.3)

$$Sh = k_l d_B/D_l \tag{9.15}$$

The value of k_l is obtained, and also the value of $k_l a$ is obtained.

For the gas side mass transfer coefficient k_g, the same estimation method with $Sh = 2$ is applied and now inserting the gas phase diffusion coefficient value. With these conservative low values of a, k_l, and k_g, the condition of no mass transfer limitation is assessed.

If the condition of no mass transfer limitation is determined, then the bubble column is a reactor type option, regarding mass transfer.

Liquid residence time distribution bubble column

If staging of the liquid flow through the reactor is required, then it can be designed as a so-called horizontal crossflow bubble column with vertical baffles to ensure liquid flow staging. Section 9.3 discusses such a design.

Heat exchange area bubble column

Heat exchange is feasible in a bubble column by installing tube bundles inside the reactor. For a concept design, the required heat exchange area can be estimated from a heat balance involving the reaction heat.

The heat exchange coefficient for the heat exchanger may be estimated by the power dissipation of a bubble column per unit liquid mass P_m and can be calculated from

$$P_m = v_{sup}\, g \tag{9.16}$$

The power dissipation by a bubble column is therefore limited by the superficial gas velocity that can be applied. Taking a maximum superficial gas velocity of 0.5 m/s results in power dissipation:

$$P_m = 5\,\text{W/kg}.$$

The heat exchanger heat transfer coefficient can be estimated by the correlation [4]

$$Nu = 0.57Re^{0.5}Pr^{0.33} \tag{9.17}$$

in which for the velocity in the Reynolds number the Kolmogorov eddy velocity v_e is inserted.

The expression for v_e is given by Beenackers van Swaaij [4] as

$$v_e = (P_m.d)^{0.33} \tag{9.18}$$

9.1.3.2 Mechanically stirred reactor performance phenomena

Mechanically stirred tank reactors have a higher gas liquid mass transfer performance because the power dissipation per unit mass can be far higher than for a bubble column by which smaller bubbles are created than in a bubble column, and also the turbulence near the gas bubbles create a higher mass transfer performance than a bubble column. The liquid phase is, however, back mixed. Heat exchange is feasible in this reactor type.

Mass transfer performance stirred tank reactor

For gas-liquid mass transfer in mechanically stirred tank reactors at commercial scale, correlations are available for a first estimate, and a good source is Yawalkar [5]. For a given chemical reaction case with liquid viscosities, an order of magnitude or higher viscosity than water, small scale experiments in a transparent reactor may be executed to determine bubble sizes. Then using the same estimate of bubble columns an estimate of the specific surface area, and the liquid side and the gas side mass transfer coefficients can be determined.

With that information, the design condition for no mass transfer limitation can be determined. If that seems feasible, then the mechanically stirred tank reactor type is an option regarding mass transfer.

Liquid phase residence time distribution stirred tank reactor

If the reaction kinetics reveal that a narrow residence time distribution is desired, then the stirred tank reactor is not an option, as its RTD behavior is fully back mixed. If a few stages are needed, then a multistage stirred reactor with multiple impellors is an option.

Heat exchange stirred tank reactor

Heat exchange in a stirred tank reactor is an option. An estimate for the required heat exchange area can be made using the same expressions based on power dissipation as those for a bubble column reactor; hence, expressions (9.17) and (9.18).

Because the power dissipation per unit mass in a stirred tank reactor can be much higher than in a bubble column, the heat transfer coefficient can also be much higher.

9.1.3.3 Static mixer reactor type

Gas-liquid static mixer reactors are used at commercial scale for many applications [6]. Here are their key performance phenomena described.

Mass transfer static mixer

The mass transfer coefficients of static mixers are in the same range as mechanically stirred tanks. The mass transfer terms $k_l a$ and $k_g a$ are governed by the turbulent field intensity parameter power dissipation per unit mass P_m, which is given by

$$P_m = v_{lsup}(dP/dL) \qquad (9.19)$$

Liquid phase residence time distribution static mixer

Liquid phase residence time distribution is narrow. A correlation for the Peclet number can be found in Chapter 21. From that information, it is clear that plug flow behavior of the liquid phase (and the gas phase) is easily obtainable.

Heat exchange static mixer

The static mixer elements can also be designed as heat exchanger tubes [6], so that heat removal (or addition) of the reaction heat is achievable. This even holds for viscous liquids. Heat exchange coefficient values for a first order of magnitude estimate can be obtained using expressions (9.17) and (9.18).

9.1.3.4 Structured packing reactor

Mass transfer

In a structured packing, reactor gas and liquid flow concurrently down the packing. The flowing liquid film is only a few millimeters thick; hence, the mass transfer coefficient k_l is high. The specific surface area per unit liquid volume is thereby also high. The $k_g a$ in structured packings reaches thereby easily up to 1 (1/s) [2].

Liquid phase residence time distribution

Plug flow of the liquid phase residence time distribution of structured packing is easily obtained, provided that the liquid distributor is properly designed. Strangely enough, I still encounter in my consultancy poorly designed liquid distributors for packed beds.

The value of the Bodenstein number Bo for the liquid flow is between 0.3–0.6 [1]. So, a conservative value is

$$Bo = 0.3$$

The Peclet number for the liquid residence time distribution is given by

$$Pe = Bo(L/d_p)$$

For a column with a height of 2 meters and a packing particle size of 0.01 m means $Pe = 60$.

So, plug flow can be easily obtained in a structured packed bed.

Heat exchange

Heat exchange in packed beds may be feasible by multitubular design. However, the radial heat transfer coefficient to the tube wall will be low, literature information is lacking, as well as industrial applications. Hence, packed bed reactors with heat exchange are not practically feasible.

9.1.3.5 Rotating packed bed reactor

Gas-liquid rotating packed bed reactors are applied at commercial scale for fast reactions requiring high mass transfer coefficients [7, 8].

Mass transfer

The mass transfer coefficients $k_l a$ and $k_g a$ in a rotating packed bed reactor are a factor 100 larger than for mechanically stirred tank reactors. Hence, this reactor is very suitable for very fast reactions.

More information on mass transfer is provided by Visscher [9].

Liquid phase residence time distribution

Hacking experimentally revealed that the liquid flow residence time distribution of a rotating packed bed is reasonable narrow [10], but so far, a generic expression for a Peclet number is not available.

Heat exchange

Heat exchange in rotating packed beds is not applicable for the same reason as for normal packed beds.

9.1.3.6 Spinning disk reactor

Mass transfer

The liquid side mass transfer coefficient $k_l a$ of rotator-stator spinning disc reactors can be 10 1/s [10]. So, mass transfer performance is high.

Liquid phase residence time distribution

The liquid phase residence time distribution of a rotor-stator spinning disk is at least equivalent to 3 back-mixed stages in series [11]. So, a 99 % conversion for fast reactions can be reached in this reactor.

Heat exchange

The heat exchange performance of a spinning disk reactor is high. The Beer measured heat transfer rates up to 34,000 W/(m^2 K) [12].

9.2 Reactor type selection method

9.2.1 Reactor selection overview

Reactor type selection for gas-liquid reactors is very important in view of:
1. Maximum product yield on feedstocks; hence, minimum feedstock cost per ton of product.
2. Minimum investment cost
3. Low scale-up risk

Table 9.1 provides an overview of 6 reactor types applied in commercial scale production and their qualitative performance characteristics. For each of these reactor types, conservative quantitative performance characteristics are provided in Section 9.1.3.

Table 9.1: Gas-liquid reactor critical performance characteristics.

Reactor	Liquid hold-up	Mass transfer g/l	Liquid RTD stages	Gas RTD stages	Heat exchange?
Bubble column	High	Low	Some	One	Yes
Stirred tank	High	Medium	Some	Some	Yes
Static mixer	High	Medium	Many	Many	Yes
Structured Packing	Low	High	Many	Many	No
Rotating packed bed	Low	High	Some	Some	No
Spinning disk	Low	High	Some	Some	Yes

The reactor type selection proposed here is as follows. Given the chemical reaction kinetics, first the required mass transfer and liquid hold-up is considered using Table 9.1 and Section 9.1.3. This selection will reduce the number of feasible and attractive reactor types. Then the liquid residence time distribution requirements are considered. Again using Table 9.1 and Section 9.1.3 feasible and attractive reactor types are reduced again. Finally, heat exchange requirements for the reactor are considered, so that finally only reactor types that fulfill all criteria for liquid hold-up, mass transfer, RTD, and heat exchange remain.

9.2.2 Reactor selection for high liquid hold-up and no mass transfer limitation

The first step in reactor selection is to select a reactor type for the given kinetics which has a high liquid hold-up and for which mass transfer is not the rate limiting step. The advantages of this approach are that:
- Reactor volume required is minimized by the high reactive liquid hold-up.
- By-product formation is minimized by no mass transfer limitations.
- Scale-up risks are greatly reduced because the reactor performance is not sensitive to mass transfer coefficient values.

Mass transfer parameter theory of how to avoid mass transfer limitations is provided in Section 9.1.2. Section 9.1.3 provides mass transfer performance of six important reactor types. Using all this information in combination with reaction kinetics a reactor type can be chosen that has no mass transfer limitation for the case at hand.

9.2.3 Reactor type selection for liquid residence time distribution

If undesired consecutive reactions of the product to by-products play a role then the liquid RTD should be narrow.

If deep conversion of a liquid feed component is desired then also the liquid RTD should be narrow.

9.2.4 Reactor type selection for heat exchange

If adiabatic reactor design is feasible then heat exchange inside the reactor is not needed. If, however, heat exchange in the reactor is needed that reactor types with heat exchange are needed. If the reaction is fast then high heat exchange capacity will be required.

9.3 Example case: Ethyl Benzene Hydro Peroxide (EBHP) reactor selection

This EBHP case has been chosen because all reactor selection aspects are relevant.

9.3.1 Chemistry EBHP

Background information on chemistry and kinetics is provided in the book on multi-phase reactors [13].

Stoichiometric equations of reactions

The desired main reaction to the product Ethyl Benzene Peroxide (EBHP)

$$\text{Ethylbenzene (EB)} + \text{Oxygen (Ox)} = \text{EBHP} \tag{9.20}$$

The undesired consecutive reaction of EBHP to by-product Methyl Phenyl Carbinol (MPC) and Propylene Oxide (PO) is:

$$\text{EB} + \text{EBHP} = \text{MPC} + \text{PO} \tag{9.21}$$

Kinetics

The kinetics of the main reaction involves a radical reaction of EB. The radical is formed from the EBHP. So, the main reaction needs some EBHP to start the reaction. So, it is autocatalytic. This reaction is sensitive to temperature.

The side reaction is a consecutive reaction. This reaction is more strongly sensitive to temperature than the main reaction.

The reaction rate is moderate with a residence time of 10 minutes, and a reasonably conversion is obtained for the reaction temperature range.

9.3.2 EBHP Reactor concept selection criteria

The commercial scale reactor capacity is in the order of 100 kt/a, The product is of high value, while the by-product (MPC) has little value. So, it is very important to have a high selectivity towards EBHP, or in other words, by-product formation should be minimized.

This means that the main reaction should be as fast as possible compared to the consecutive reaction. The main reaction is with oxygen, which is transferred from the gas phase to the liquid phase. This means that the gas-liquid mass transfer should be high compared to the reaction.

Because the side reaction is a consecutive reaction it means that the liquid flow residence time distribution should be close to plug flow.

Oxygen feed is not very expensive but the gas compressor to overcome the reactor pressure drop is expensive. So, to minimize the gas feed and utilize the oxygen as much as possible the gas flow should be plug flow or at least staged.

The heat of reaction is high at it is an organic oxidation reaction. The reaction temperature should be close to the optimized temperature for the main reaction, relative to the consecutive reaction. This means that the reactor has to be cooled.

The total required reactor liquid volume for the reactors can be derived from the picture provided in Harmsen [13] and it is at least 150 m^3.

9.3.3 EBHP reactor type options evaluation

Bubble column reactor

This reactor has liquid flow back-mixed behavior and the gas phase RTD is complex. So, this is not a desired reactor type.

Horizontal crossflow liquid staged bubble column reactor

This reactor type has many liquid flow stages due to vertical baffles. Reaction heat can be removed by placing heat exchanger tubes inside the reactor. Due to the high energy dissipation of the gas flow, reasonably high heat transfer is achieved and also a reasonably degree of liquid mixing to avoid local hot spots.

Cocurrent multitubular reactor

A cocurrent gas-liquid multitubular packed bed reactor type would have all desired reactor concept features, except for the liquid hold-up fraction, which is in the order of 30 %. So, the reactor volume of this reactor type would be three times higher than for the horizontal bubble column reactor. It also would need a back-mixed reactor up front of this plug flow reactor because the main reaction is autocatalytic by the product.

Mechanically stirred reactor

This reactor type has in residence time distribution terms, back-mixed liquid flow behavior. So, it is not desired in view of minimizing the consecutive reaction and also for deep conversion in a limited volume.

9.3.4 Reactor scale-up method

The mass transfer, and the gas and liquid residence time distribution and back-mixing of horizontal crossflow baffled bubble columns can be studied by a cold flow model. Some major scale-up uncertainties can so be removed. This has indeed been applied for the application of this reactor type.

9.3.5 Reactor selection evaluation

Shell has operated this EBHP reactor on a commercial scale since the 1970s of the last century. In the first decade of this century, the reactor gas distribution and baffles have been optimized by CFD modeling and tracer studies at the commercial scale reactors. Proving that the original reactor design was suboptimal but good enough for commercial scale exploitation [13].

Bibliography

[1] Westerterp KR, Van Swaaij WP, Beenackers AA. Chemical Reactor Design and Operation. 1984.

[2] Stankiewicz A. The principles and domains of process intensification. Chemical Engineering Progress. 2020 Mar 1;116(3):23–8.

[3] Li P, Zhu DZ, Wang H, Tang R. Methods for predicting bubble size distribution in turbulent flow. Water Resources Research. 2025 Apr;61(4):e2024WR038386.

[4] Beenackers AA, Van Swaaij WP. Slurry reactors, fundamentals and applications. In: Chemical Reactor Design and Technology: Overview of the New Developments of Energy and Petrochemical Reactor Technologies. Projections for the 90's. Dordrecht: Springer Netherlands. 1986. pp. 463–538.

[5] Yawalkar AA, Heesink AB, Versteeg GF, Pangarkar VG. Gas—Liquid mass transfer coefficient in stirred tank reactors. The Canadian Journal of Chemical Engineering. 2002 Oct;80(5):840–8.

[6] Thakur RK, Vial C, Nigam KD, Nauman EB, Djelveh G. Static mixers in the process industries—A review. Chemical Engineering Research and Design. 2003 Aug 1;81(7):787–826.

[7] Trent DL. Chemical processing in high-gravity fields. In: Re-Engineering the Chemical Processing Plant. CRC Press; 2018 Dec 14. pp. 45–79.

[8] Pei DY, Su MJ, Wang YY, Chu GW, Luo Y, Chen JF. Process intensification of 2,3,6-trimethylphenol oxidation in a rotating packed bed reactor. Chemical Engineering and Processing-Process Intensification. 2020 Mar 1;149:107842.

[9] Visscher F. Van der Schaaf J, Nijhuis TA, Schouten JC. Rotating reactors—A review. Chemical Engineering Research and Design. 2013 Oct 1;91(10):1923–40.

[10] Hacking JA, Delsing NFEJ, de Beer MM, van der Schaaf J. Improving liquid distribution in a rotating packed bed. Chemical Engineering and Processing: Process Intensification. 2020;149:Article 107861. https://doi.org/10.1016/j.cep.2020.107861.

[11] Visscher F, de Hullu J, de Croon MH, van der Schaaf J, Schouten JC. Residence time distribution in a single-phase rotor–stator spinning disk reactor. AIChE Journal. 2013 Jul;59(7):2686–93.

[12] De Beer MM, Keurentjes JT, Schouten JC, van der Schaaf J. Intensification of convective heat transfer in a stator–rotor–stator spinning disc reactor. AIChE Journal. 2015 Jul;61(7):2307–18.

[13] Harmsen J, Bos R. Multiphase Reactors: Reaction Engineering Concepts, Selection, and Industrial Applications. De Gruyter; 2023.

10 Two-phase heterogeneously catalyzed reactors

10.1 Basics reactor type selection

Many reaction systems are catalyzed by porous heterogeneous catalysts. Two very different reactor families are suitable for heterogeneous catalysts: fixed bed reactors and fluid bed reactors.

Fixed bed reactors are in general easy to design, and reliable scale-up methods have been established. Their residence time distribution can be designed to be plug-flow, or if back-mixing is desired, by involving a recycling flow. Reliable mass transfer expressions are available for sizing and shaping catalyst particles. Heat exchange is feasible by having multiple tube design; however, this is an expensive design solution. Catalyst replacement can only be done by shutting down the operation.

Fluid bed reactors are harder to design reliably. Their "fluidized medium," fluid and solids, has a wide residence time distribution and is for specific design types not easy to predict for commercial scale reactors.

Their heat transfer coefficients, however, are much higher than for fixed bed reactors. So, for that reason, fluid beds are attractive if heat exchange is needed.

If the heterogeneous catalyst is quickly fouled by coke deposition, then fluid bed reactors are also superior to a fixed bed reactor, as the coked catalyst can be withdrawn from the reactor under fluidization conditions, be regenerated by oxidation in a second fluid bed reactor, where the coke is burned off, and the regenerated catalyst can then be fed back to the main reactor. Fluid Catalytic Cracking (FCC) is a prime example of this set of reactors.

For adiabatic reactors, a fixed bed is for most cases the preferred first option to consider. The effect of size and shape of these catalysts on diffusion limitations are first to be analyzed by reaction engineering methods. If the particle size for the absence of diffusion limitation is larger than 0.5 mm, then in general a fixed bed reactor can be selected.

If, however, the particle size needs to be smaller than 0.5 mm then a fluid bed reactor may be selected. This is of course a too simple criterion for a decision between a fixed bed and a fluid bed reactor. Diffusion limitations can also be avoided by selecting special particle shapes such as rings or trilobes so that a packed bed reactor can still be used.

10.1.1 Catalyst particle size design avoiding internal diffusion limitation

Heterogeneous catalysts are the most commonly used catalysts in industry. Nearly all those catalysts are porous catalyst particles. These catalysts have a large catalytic pore surface area, of several hundred of square meters per gram of catalyst particles. Molecules diffuse through the pores and react on the pore surfaces to the desired product. The product diffuses through the pores to the catalyst particle outside. If the reaction

https://doi.org/10.1515/9783111203256-014

is fast relative to the diffusion, then the concentration of the reacting molecules drops along the pores from the outside to the inside. This in general slows down the reaction rate so the catalyst is less effective than the theoretical maximum. The diffusion rate depends on the diffusion length. By reducing the particle diameter, the diffusion length is reduced; hence, the catalyst is then more effective.

If the product reacts further to an undesired by-product and the reaction rate is in the same order o magnitude as the main reaction, then diffusion limitation is even less desired. As the diffusion of the product out of the catalyst particle is also slower than the formation rate, the product concentration inside the catalyst is higher than outside. This means that the undesired reaction rate is higher than when no diffusion limitation is present.

The reaction rate relative to the diffusion rate is described by the Thiele Modulus ϕ [1].

It is a dimensionless number and represents the ratio of the reaction rate and the diffusion rate.

For reaction rates:

$$-r_A = kC_A^n, \tag{10.1}$$

in which r_A is defined a mole reacted per catalyst particles volume (m^3) per second

C_A = is concentration of A moles/m^3

k = reaction rate constant $(mol.m^3)^{n-1}\,s^{-1}$

D_{diffef} = the effective diffusion coefficient for species A inside the catalyst particle

The Thiele modulus ϕ [1] is given by (10.2):

$$\Phi^2 = (L^2/2D_{eff}).(n+1)k/C_A^{n-1} \tag{10.2}$$

The pore length parameter L is defined by

$$L = \text{particle volume/exterior particle surface [2].}$$

Table 10.1 contains some expressions for calculating L.

Table 10.1: Characteristic particle length dimension L for diffusion [2].

Particle shape	L
Sphere	$d_p/6$
Cylinder	$d_p/4$
Flat plate	$d_p/2$
General	Particle volume/external surface

The effectiveness of the catalyst is now uniquely related to the Thiele modulus. If the Thiele modulus is less than 0.2, then the effectiveness is 1, meaning that the concentration of A is uniform inside the pores and diffusion limitation is absent.

With this information, the optimum catalyst shape and size can now be determined for maximum catalyst effectiveness, by choosing the L value such that $\phi = 0.2$ or choosing an even lower value for L, so that if by catalyst optimization higher reaction rates are obtained, still no diffusion limitation occurs.

When performing experiments to determine reaction kinetics, it is advisable to also vary the catalyst particle size for the highest reaction rates to confirm that diffusion limitations are absent.

The catalyst shape and size should not only be designed for effectiveness, but also for catalyst particle strength and pressure drop when applied in fixed bed reactor applications. This subject is treated in Section 10.2.

When very fine catalyst particles are desired to avoid diffusion limitations, i. e., when particles are to be smaller than 0.5 mm to avoid diffusion limitation then a fluid bed reactor is in general preferred. Some aspects of fluid bed reactor design are treated in Section 10.2.3.

10.1.2 External mass transfer limitation evaluation

Mass transfer limitation outside the porous catalyst particles is in general also absent when there is no internal diffusion limitation. The reasoning to arrive at this conclusion is that the thickness of the external boundary layer ∂ for mass transfer is maximally $0.5\,d_p$. Hence, it is the same diffusion length as for the internal diffusion. This maximum boundary layer thickness follows directly from the minimum value of the Sherwood number Sh = 2, which holds for very low flow velocities around the particle.

Moreover, the external diffusion coefficient is always higher than the internal diffusion coefficient. Hence, if there is no internal diffusion limitation, then there is also no external diffusion limitation, regardless of how slow the gas or liquid flows around the catalyst particles.

If, however, internal mass transfer limitations are present, then external mass transfer limitations may be present. External mass transfer coefficients can be calculated from the Sherwood number using correlations given in Section 10.2.3.

10.2 Reactor design method

10.2.1 Reactor design introduction

In this section, two reactor types are treated: packed beds and fluid beds. Major design aspects are described, and also scale-up and scale-down methods are briefly described

10.2.2 Packed bed reactor design

The main items for packed bed reactor design are residence distribution, mass transfer, and pressure drop. These design items are treated both for lab scale design and commercial scale design.

10.2.2.1 Design for narrow residence time distribution

For deep conversion in a limited reactor volume, a narrow residence time distribution is desired so that plug flow is obtained resulting in a minimum reactor volume. The relation between conversion and the required residence time distribution to obtain plug flow is given by the minimum Peclet number Pe_{min} reported by Mears [3]. Details are provided in Chapter 8:

$$Pe_{min} = 8n \ln(1/(1 - X)) \tag{10.3}$$

in which n is the reaction order in the reactant (reactant $\rightarrow$ product) and X is the conversion degree.

For packed beds, the Peclet number Pe is defined as

$$Pe = v_{sup}L/D_{ax} \tag{10.4}$$

$$V_{sup} = \text{the superficial fluid velocity (m/s)}$$

$$L = \text{the bed length (m)}$$

$$D_{ax} = \text{the axial dispersion coefficient (m}^2\text{/s)}$$

For a packed bed, the Peclet number can be determined from

$$Pe = (L/d_p)\,Bo, \tag{10.5}$$

in which Bo is the Bodenstein number:

$$Bo = v_{sup}d_p/D_{ax} \tag{10.6}$$

For Reynolds number, Re > 10, Bo = 2 [4].

For Re $\ll$ 1, no Bo correlations for fluid-particle flow are available. A conservative value for liquid in trickle flow provided by Gierman [5] can be used:

$$Bo = 0.01.$$

The Reynolds number for packed beds is given by

$$Re = \rho v_{sp}d_p/\eta. \tag{10.7}$$

So, by choosing the linear velocity such at Re > 2, the Bodenstein number = 2, and the packed bed reactor length can be defined for obtaining plug flow. For scaled down fixed bed, the conservative Bo = 0.01 can be used.

10.2.2.2 Design to avoid fluid flow mass transfer to particles

For fluids (gas or liquid) passing through a packed bed, the mass transfer coefficient k_1 can be determined from expression (10.8) from Ranz [6].

$$Sh = 2 + 1.8\,Re^{0.5}\,Sc^{0.33} \tag{10.8}$$

Mass transfer for a single particle is given by Froessling [7]:

$$Sh = 2 + 0.6\,Re^{0.5}\,Sc^{0.33} \tag{10.9}$$

The higher mass transfer coefficient for a packed bed compared to a single particle is that the actual fluid velocity in the packed bed is higher than the superficial velocity, while for a single particle the actual velocity is the same as the superficial velocity.

The packed bed expression has no correction for the void fraction. For void fractions far higher than 0.4, it will be unreliable.

10.2.2.3 Bed length h in view of pressure drop and catalyst breakage

The pressure drop and the weight of the catalyst bed exerts a force on the packed bed particles. The maximum force is at the bottom of the packed bed when the fluid flows downwards. If the force on the bottom particles exceeds the bulk crushing strength, then the catalyst particles break down.

The catalyst bulk crushing strength at the bottom the catalyst bed should not be superseded. So, the packed bed gravity and the pressure drop together should be less than this bulk crushing strength. This is a conservative design method as part down forces will be counteracted by the forces on the reactor wall.

This effect is a function of the friction coefficient and of the bed length to vessel diameter ratio. For narrow tubes with bed length/bed diameter $\gg 1$, such as in a multitubular packed bed reactor, the gravity force and pressure drop force on the bottom particles is small as nearly all forces are counteracted by the wall friction force.

The fluid flow pressure drop has to be added to the gravity force. The pressure drop over a packed bed can be calculated with the Ergun correlation [8].

$$\left(\frac{\Delta P}{\rho_f u_s^2}\right)\left(\frac{d_s}{L}\right)\left(\frac{\varepsilon_b^3}{1-\varepsilon_b}\right) = 150\frac{1-\varepsilon_b}{Re} + 1.75 \tag{10.10}$$

Where ρ_f is the fluid density, u_s the superficial velocity 10 and L the bed length. For the Reynolds number different equivalent diameter should be used, depending on the particle shape, see Table 10.2.

Table 10.2: Equivalent hydraulic diameters for particles of different shape for pressure drop.

Shape	Equivalent diameter (d_p)
Sphere, diameter d_s	d_s
Cylinder with length H equal to diameter $d_c = 2R_u$	d_c
Long extrudates with radius R_u	$3R_u$
Ring with length H, outside radius R_u, and inside diameter R_i	$\frac{3(R_u + R_i)H}{R_u - R_i + H}$

The Ergun expression can be used for engineering estimates (±30 %) for pressure drop in fixed beds of (catalyst) particles with regular shapes like spheres and cylinders, provided an appropriate hydraulic or equivalent diameter and shape factor is applied.

10.2.2.4 Scaled down packed bed design

For cases where the kinetics are not simple, catalyst poisoning is expected to play a major role, or for cases where ion exchange catalysts are used, a scaled-down mini-reactor will be needed to validate the complex kinetics, to obtain reliable catalyst poisoning rates, or ion exchange resin activity bed profile.

The scale-down rule of such a fixed bed for avoiding wall flow can be borrowed from the scale-down rule for three-phase fixed bed reactors; $D/d_p > 25$, in which D is the fixed bed diameter and d_p the particle diameter. Details are presented in Section 25.3.

10.2.3 Gas-solid fluid bed reactor design

10.2.3.1 Particle size selection for fluid beds

Adiabatic fluid bed gas-solid reactor concept design is a subject with many aspects to be considered. First of all, the residence time distribution of the gas and of the solids is wide and depends among others on particle size distribution, reactor dimensions, and gas velocity. The main reasons for choosing a fluid bed reactor over a fixed bed reactor are that solids can be easily processed in continuous operations, fine particles can be processed, and the reactor construction cost can be low compared to fixed beds.

The particle size distribution and the densities of gas and solids play a major role in the fluid flow behavior of the fluid bed. Figure 10.1 is the so-called Geldart diagram [9]. It shows four fluid bed-class behaviors: Class C; cohesive, Class A; aeratable, Class B; sand-like and Class D; spoutable, for catalytic fluid beds class A is the preferred class. In class A fluid beds, the fluid behaves like a liquid with a low viscosity. Addition of solids and removal is easy. Heat transfer to a heat exchanger is high. Class A behavior can be obtained by choosing the particle size based on Figure 10.1.

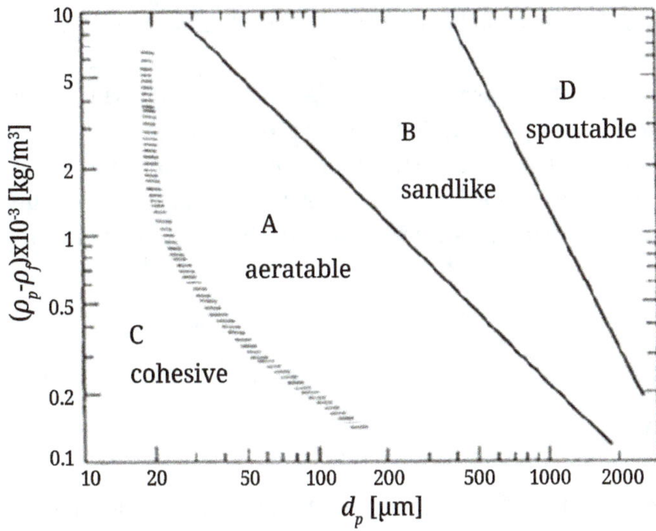

Figure 10.1: Geldart diagram Fluid bed behavior classes [9].

10.2.3.2 Fluid bed solids flow residence time distribution by design

The solids residence time distribution of a single fluid bed is very wide, and accurate predictions are hard to make. There are. However. three known methods to control the RTD by design.

One option is to have a large external solids flow recycle so that the RTD can be considered back mixed.

The second option is to have several reactors in series so that a minimum of solids staging is obtained by the number of reactors.

The third is to have crossflow of gas and solids in a horizontal fluid bed with compartments. The solids flow horizontally from one compartment to the next so that staging of the solids residence time distribution is obtained. A textbook on gas-solid fluid beds is provided by Kunii [10].

10.2.3.3 Gas residence time distribution by design

The residence time distribution of the gas phase through a fluid bed is very wide. Part of the gas flows quickly as "bubbles" through the bed and another part flows slowly via the "dense" phase through the bed. The gas phase residence time distribution is furthermore strongly affected by the vessel dimensions.

Moreover, the gas phase RTD is strongly affected by the particle size distribution. Most academic studies are carried out with particle size distributions in Geldart classification B, while commercial scale applications are mostly for Geldart classifications A. The fluid bed behavior of class A is very different from class B behavior, as the dense phase effective viscosity is a factor 100 or more lower than for class B. This means that experimental results of class B and the models based on these results have little meaning for class A fluid bed behavior.

There is, however, a design that avoids uncertainties about the gas phase RTD and that is a design with a large external gas recycle to gas feed ratio. Typically, a ratio of 20–100. By this, the gas phase residence time distribution is close to fully back mixed, and the gas phase concentrations inside the fluid bed are nearly uniform. By this, quality control is enabled and scale-up uncertainty is greatly reduced. This design is applied for polyolefin fluid bed reactors [11].

10.3 Example case: Industrial BPA ion exchange catalyst reactor

10.3.1 Introduction to BPA case

Bisphenol A formation from acetone and phenol using Amberlyst 15 ion exchange resin particles as a heterogeneous acid catalyst has been researched experimentally by Lilian, for her MSc thesis [12]. This case is taken to show how preliminary experimental results of the ideation stage can be used to make a first commercial scale reactor concept design, which is then used to guide for research in the concept stage.

Lilian reports her experimental results with no chemical reaction engineering analysis on kinetics. So, preliminary kinetics with an order of the reaction is assumed and kinetic parameters are estimated from the preliminary batch experiments. The Thiele Modulus value is then estimated to design a particle size followed by a fixed bed commercial scale concept design. From that a downscaled mini-plant fixed bed reactor is designed for experimental validation.

10.3.2 BPA chemistry and kinetics

The reaction stoichiometry for the desired reaction to BPA is

$$\text{Acetone} + 2\text{Phenol} = \text{BPA} + \text{water}$$

The acid catalyzed reaction mechanism is via an intermediate

$$\text{Ac} + \text{Ph} + \text{H}^+ \rightarrow \text{I}$$
$$\text{I} + \text{Ph} \rightarrow \text{BPA}$$

Lilian performed an experiment in a batch reactor with 10 % Amberlyst in the phenol acetone mixture. After 5 hours, the yield of BPA was 0.4 % [12]. The first-order reaction rate taking into account the 10 % of catalyst in the reaction fluid, results in first-order rate being constant per unit catalyst volume:

$$k_1 = 1.7\,10^{-3}\,\text{s}^{-1}$$

10.3.3 BPA ion-exchange catalyst particle Thiele modulus analysis and particle size design

The Thiele modulus value was determined by assuming that the diffusion coefficient for acetone in the ion exchange resin is 10^{-10} m^2/s. The particle diameter of Amberlyst 15 has range of 0.3–1 (10^{-3} m). The largest value of this range was chosen; hence, $d_p = 10^{-3}$ m.

The resin particle is a sphere so $L = d_p/6$.

With the first-order reaction in acetone and with the reaction rate constant value $k_1 = 1.7\,10^{-3}\,\text{s}^{-1}$, this results in

$$\Phi = 0.5.$$

This means diffusion inside these largest particles may have some diffusion limitation. By choosing a smaller particle size or reducing the reaction rate by a different ion exchange resin, or by lowering temperature, this limitation can be overcome.

If there is no consecutive reaction of the product to a by-product, then this diffusion limitation is not a large problem. It just means that the reactor volume required is larger than when there would be no diffusion limitation. If, however, a consecutive reaction takes place, then diffusion limitation is not desired as it means that the product concentration inside the particle is larger than in the bulk of the liquid, Wwhich means that the consecutive reaction is faster than when diffusion limitation would not be present. So, then a smaller particle size should be chosen, or the temperature should be lower to reduce the reaction rate.

10.3.4 BPA commercial fixed bed concept design for plug-flow

For the whole process, deep conversion of acetone is desired to avoid acetone recycle. Acetone recycle involves acetone separation from the product by distillation, so an additional process step.

So, plug flow for deep acetone conversion is desired.

Let us assume an acetone conversion of 99.99 %. So $1 - X = 0.0001$.

With the first-order reaction rate for acetone determined in the section above and with a catalyst volumetric holdup of 60 %, it means that a residence time of 9,000 s is needed.

The maximum bed length for the Amberlyst particle to avoid particle crushing is not known. We assume a maximum bed length of 7 m. This means that the superficial liquid velocity = $0.8\,10^{-3}$ m/s. This results in a particle Reynolds number:

$$Re = 0.2$$

For this low Reynolds number, the conservative value Bo = 0.01 is used.

With Pe = Bo L/d_p, this results in Pe = 350.

The minimum required Peclet number for plug flow behavior for the 99.99 % conversion using expression 10.3 results in Pe_{min} = 74.

So. plug flow of the commercial scale for this deep conversion is obtainable.

10.3.5 Concept stage research plan with mini plant fixed bed reactor

The research plan for the BPA reactor should contain the following.

A kinetics study

In the kinetics study, experiments should be carried out to determine the kinetics of the main reaction as the kinetics of by-products formed as function of concentrations of acetone and BPA and temperature.

The kinetic information should also include the effect of ion exchange resin dehydration in the first days in the presence in the phenol. The water produced by the reaction will also affect the ion exchange acidity; hence, its activity. This water formed by the reaction will affect the catalyst activity locally so that the catalyst will have an activity profile over the bed length. This activity should be properly kinetically modeled and be validated by mini-bed experiments.

A mini-fixed bed reactor design

With the kinetic information obtained, an optimized commercial scale fixed bed reactor should be designed.

Then also downscaled mini-fixed bed reactor should be designed to study catalyst degradation and confirm the complex kinetics with a water content profile over the bed length for the fluid and the ion exchange catalyst.

The small-scale fixed bed needs to have the same residence time (same liquid hourly space velocity), sufficient plug flow behavior, and no preferential wall flow.

The main design rules are then:

Avoid wall flow by D/d_p > 25. So, the bed diameter should be 2.5 cm, or wider.

To have plug flow behavior for deep acetone conversion of 99.99 %, the Peclet number should be Pe > 74. With Bo = 0.02, this means that L/d_p = 3,700. So, the bed should be 3.7 m long.

The liquid flow residence time distribution can be determined in the mini-fixed bed reactor by pulse injection and outlet response recording. The pulse injection can be a salt, or fluorescent material. In this way, a more accurate residence time distribution behavior of the small-scale fixed bed will be known.

This mini-fixed bed can then be run for a long time to determine the catalyst life-time. In addition, a temperature operating range can be established. This all can lead to an optimized commercial scale reactor design.

Bibliography

[1] Thiele EW. Relation between catalytic activity and size of particle. Industrial & Engineering Chemistry. 1939 Jul;31(7):916–20.
[2] Levenspiel O. Chemical Reaction Engineering. Hoboken: John Wiley & Sons; 1998.
[3] Mears DE. The role of axial dispersion in trickle-flow laboratory reactors. Chemical Engineering Science. 1971;26:1361.
[4] Harmsen J, Bos R. Multiphase Reactors: Reaction Engineering Concepts, Selection, and Industrial Applications. De Gruyter; 2023.
[5] Gierman H. Design of laboratory hydrotreating reactors: Scaling down of trickle-flow reactors. Applied Catalysis. 1988 Jan 1;43(2):277–86.
[6] Ranz WE. Friction and transfer coefficients for single particles and packed beds. Chemical Engineering Progress. 1952;48:247–53.
[7] Froessling N. Gerland Beitr. Geophys. 1938;52:170. Elgeti K. A new equation for correlating a pipe flow reactor with a cascade of mixed reactors. Chemical engineering science. 1996 Dec 1;51(23):5077–80.
[8] Ergun S. Fluid flow through packed columns. Chemical Engineering Progress. 1952;48:89–94.
[9] Geldart D. Gas Fluidization Technology. Chichester: John Wiley and Sons; 1986.
[10] Kunii D, Levenspiel O. Fluidization Engineering. Elsevier; 2013 Oct 22.
[11] Soares JB, McKenna TF. Polyolefin reaction engineering. Weinheim, Germany: Wiley-VCH; 2012.
[12] Neagu L. Synthesis of Bisphenol A with Heterogeneous Catalyst. Master of Science theses, Queen's University, Kingston, Ontario, Canada, August 1998.

11 Three-phase heterogeneous catalyst reactors

11.1 Basics three-phase reactors

11.1.1 Introduction

In three-phase catalytic reactors, the reaction takes place inside a porous heterogeneous catalyst. Some of the reacting components are supplied as liquid to the reactor and some components are supplied as gas to the reactor. This means that all these components have to travel from these phases to the inside of the porous particles. The product components travel from porous particles to the liquid and the gas phases. Due to all these mass transfer phenomena, designing these reactors are complex.

This complexity is increased because in many cases the residence time distributions of gas and liquid are also critical performance phenomena, when deep conversion is required, or when undesired consecutive reactions occur.

Chemical reactions also generate heat or require heat. This heat has to be managed by design. In some cases, the design solution can be adiabatic operation, but in many cases heat exchange in the reactor has to take place.

Catalysts degrade by poisoning. This can be by reversible fouling, or by active sites loss. Reversible fouling can be counteracted by regeneration as part of the reactor system. Poisoning cannot be counteracted by regeneration but still needs catalyst withdrawal from the reactor. For some reactor types, this is feasible under operation.

11.1.2 Three phase reactor types

The main three-phase reactor types at commercial scale are trickle-flow fixed bed, bubble column, mechanically stirred tank, and oscillating flow reactor. The latter three are slurry reactors.

All four reactor types can have different configurations with different performance characteristics. Therefore, each reactor type is described in this section on their main characteristics. In Section 11.2, their selection and design methods are described.

11.1.2.1 Trickle-flow fixed bed reactor

In the trickle bed, reactor gas and liquid flow cocurrently downwards over the catalyst particles. Mass transfer of gas components and liquid components to the catalyst particles is in general high and often external mass transfer limitations are absent for a well-designed reactor.

Heat removal can be by designing the reactors as multitubular. Each tube wall acts then as heat exchange area.

https://doi.org/10.1515/9783111203256-015

Heat removal can also be obtained by applying a large surplus of gas to the reactor, as the gas flows cocurrently with the liquid there is no flooding limitation.

11.1.2.2 Three-phase bubble column slurry reactor

The simplest configuration of a three-phase column reactor is a cylindrical vertical column. It is cheap to manufacture, so it is the first reactor type to consider. Gas is fed from the bottom via a sparger. For low superficial gas velocities, typically, with $v_{gs} < 0.1\,\mathrm{m/s}$, gas bubbles are in the so-called homogeneous regime with small differences between bubble sizes.

For higher gas velocities, gas flows partly as big bubbles fast in up flow through the column. Small bubbles are also present, which flow upwards in the center and downwards near the wall. Liquid flows fast up flow through the center and in fast down near the wall. The fine catalyst particles follow these liquid flows, with hardly any difference in velocity with the liquid. This means that the three-phase bubble column is similar in behavior for gas-liquid mass transfer and residence time distribution to the two-phase bubble column described in Chapter 10. This means that the gas residence time distribution is wide.

If the reactor is operated batchwise for the liquid phase, then the liquid residence time distribution is plug flow. The same then holds for the catalyst RTD. If the reactor is operated continuously for the liquid flow, then the liquid flow will have a large residence time distribution, the same as described in Chapter 10 for two-phase bubble columns.

The external mass transfer coefficient k_l can be approximated with Sh = 2; hence, $k_l = 2D/d_p$, because the slip velocity between solids and liquid is small.

Heat removal can be obtained by placing heat exchange tubes inside reactor.

11.1.2.3 Mechanically stirred tank slurry reactor

The three-phase mechanically stirred reactor has higher gas-liquid mass transfer coefficients than the bubble column. So, if gas-liquid mass transfer is the limiting rate in the bubble-column, then a mechanically stirred tank can be considered.

The liquid phase residence time distribution is plug flow if the reactor is operated in batch mode. If it is continuously operated, then the liquid phase will be back mixed. The gas flow residence time distribution will in general be wide.

Heat removal can be designed by placing heat exchange tubes inside the reactor.

11.1.2.4 Oscillatory flow baffled slurry reactor

In the oscillatory flow baffled reactor, the fluids flow through a tube with baffles at the wall. Superimposed on the flow is a periodic forth and backward flow which creates in combination with the baffles a turbulent field. The flow has near plug-flow behavior for liquid, solids, and gas flow [1].

Because the reverse flow intensification can be high, the gas-liquid mass transfer can be high. The mass transfer is decoupled from the average liquid and gas velocity and, therefore, extremely low average velocities can be chosen, so long residence times can be applied in a limited reactor length.

The reactor wall can act as a heat exchanger. The turbulent field creates high heat transfer coefficients.

Because of all these features, it is a versatile reactor type especially for slow reactions.

11.2 Reactor-type selection method

11.2.1 Reactor-type selection overview

Three-phase heterogeneous catalyzed reactors can be divided into two main categories, fixed bed, and slurry reactors. Key differences between the two categories are catalyst particle size, residence time distribution of gas and liquid, heat transfer performance, ease of catalyst replacement, and scale-up risk, as shown in Table 11.1.

Table 11.1: Key differences between three-phase fixed bed and slurry reactors.

Performance phenomenon	Trickle bed	Slurry
Catalyst regeneration in process	No	Yes
Mass transfer g/l	High	Medium
Mass transfer l/s	Medium	High
Mass transfer by diffusion	Low	High
Residence time distribution gas	Narrow	
Residence time distribution liquid	Narrow	
Heat transfer	Adiabatic by surplus gas	Heat exchanger inside
	Multitubular heat exchange	

11.2.2 Reactor selection for catalyst regeneration

Heterogeneous catalysts in general degrade in time. This degradation can be reversible or irreversible. The most common reversible catalyst degradation is by coke deposition. The coke can be removed by oxidation with oxygen. The reactor type selection depends on the stream time in between regenerations as shown in Figure 11.1 [2].

For stream time longer than 1 year, a packed bed can be chosen. Each time catalyst regeneration is needed the reactor is shut down and catalyst is regenerated. For much shorter stream time, multiple reactors will be designed, so that one reactor can be shut down for regeneration, while the other reactors are kept in operation.

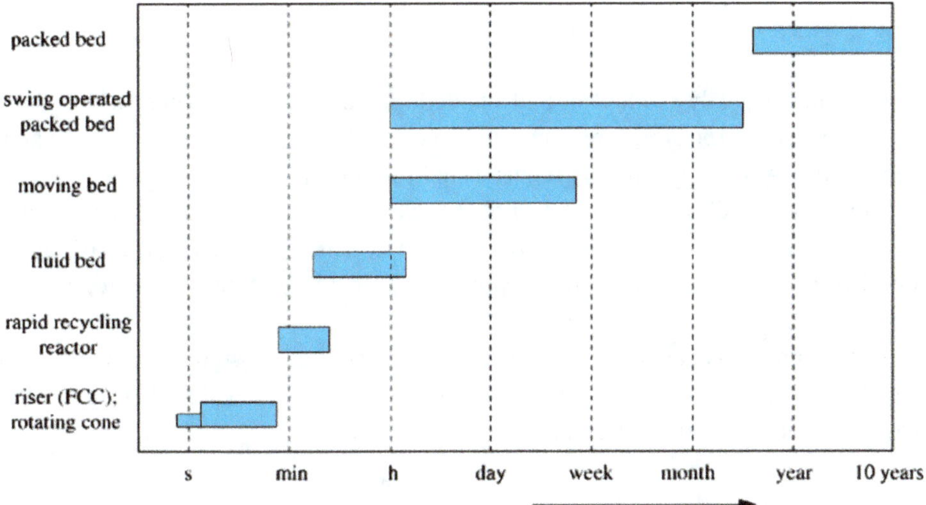

Figure 11.1: Reactor type applied versus on stream time catalyst.

For short stream times, a slurry reactor is preferred.

11.2.3 Reactor selection for high mass transfer

If reactions are fast compared to diffusion inside the catalyst particles, then small particles are desired and then a slurry reactor is the preferred option.

11.2.4 Reactor selection for residence time distribution liquid

If a narrow liquid residence time distribution is desired for high product selectivity and deep conversion in a limited reactor volume, then a fixed bed reactor is preferred.

11.2.5 Reactor selection for residence time distribution gas

If a narrow gas residence time distribution is desired for deep gas component conversion in a limited reactor volume, then a fixed bed reactor is preferred.

11.2.6 Reactor selection for heat exchange

If the reaction is highly exothermic and adiabatic and the operation is not feasible, then a slurry reactor is preferred over a fixed bed in view of investment cost, as heat transfer coefficients in a slurry reactor are higher than in a multitubular fixed reactor.

However, this difference is often not the overriding factor, as heat transfer only affects the investment cost, while catalyst particle size, and residence time distribution affect by-product formation and thereby product yield on selectivity. These are in most cases overriding cost contributions.

For gas-liquid fixed bed reactors operating in gas-liquid down flow, Taulamet performed an experimental study on heat transfer of tubular trickle-bed reactors for a wide range of particles and flow rates [3].

11.2.7 Three-phase fixed bed reactor design

Three-phase fixed bed reactors are nearly always designed for cocurrent down flow of gas and liquid, as this ensures a large design window for gas and liquid velocities with no danger of fluidization, which is the case for up flow. So, the remainder of this section is about cocurrent down flow fixed bed reactors, called trickle-bed or trickle-flow reactors.

11.2.7.1 Three-phase fixed bed design for narrow residence time distribution

For trickle-bed reactors at commercial scale, a narrow liquid residence time distribution can be obtained from empirical correlations of the Peclet number versus the Reynolds number. Stegeman summarized a number of literature correlations into one correlation [4]. For Re > 10, Bodenstein appears to be at least 0.4. Hence, a conservative expression is

$$Pe = 0.4L/d_p$$

For a commercial scale bed with L = 1 m, Peclet is already 40, so plug flow is obtainable most for design cases.

The liquid RTD also depends strongly on the uniform liquid distribution over the packed bed. Gas and liquid distributor design is part of the EPC stage, so not part of the design focus of this book. However, because in my career I have noticed several cases of too few distributor points for liquid distribution this paragraph is added.

For multitubular trickle-bed reactors, it is of utmost importance to have an even distribution of gas and liquid over the individual tubes. Special gas-liquid distribution design methods are available, and Maiti provides a review of the subject [5].

11.2.7.2 Three-phase fixed bed design for mass transfer

Both gas-liquid mass transfer coefficients k_g and k_l of three-phase fixed bed can conservatively be estimated by

$$Sh = k \, d_p/D = 2$$

in which k can be replaced by k_g and by k_l and D by the gas phase diffusion coefficient and the liquid phase diffusion coefficient.

The liquid layer thickness flowing of the particle is less than the particle diameter so the specific surface area a can be conservatively estimated by

$$a = 1/d_p.$$

For most industrial cases, mass transfer limitations of the gas and liquid side will not occur as long as diffusion limitations inside the porous catalyst particles also do not occur.

11.2.8 Three-phase heterogeneous catalyst slurry reactor design

Three-phase slurry reactor design is far more complex than fixed bed design and is therefore beyond the scope of this book. The reader is referred to textbooks on this subject. An entrance is multi-phase reactors [6]. Section 11.3 includes a three-phase slurry reactor.

11.3 Example case: Reactor selection HydroDeSulfurization and HydroDemetallization

11.3.1 Introduction to HDS and HDM

This case is about Hydro DeSulfurization (HDS) and Hydro DeMetallization (HDM) reactor type selection in view of all criteria including catalyst lifetime as this the main subject of Section 11.2. The design of a commercial scale HDS reactor is provided in Section 25.3; so, it is not treated here.

The reaction of many organic sulfur components to be hydrogenated can be lumped into one reaction stoichiometry:

$$H_2(g) + R\text{-}SH(l) = RH(l) + H_2S(g) \tag{11.1}$$

The reactor operates typically at 150 bar hydrogen to avoid coke formation in the porous catalyst.

The reaction of organic sulfur metal components with hydrogen is here simplified to

$$H_2 + MSCH = RH(l) + MS(s) \tag{11.2}$$

The resulting metal component deposits in the porous catalyst, so that the catalyst becomes inactive.

The reaction temperature is in the range 350–450 °C.

All individual reactions are first order in R-S-H components. But because some components react faster than other components, the overall lumped kinetics show a higher-order effect in R-SH concentrations. The reactions are furthermore second order in hydrogen concentrations. Organic metal components react at a lower rate than the sulfur components.

All hydrogenation reaction heats for our concept design purpose can be lumped into $\Delta H = -40\,kJ/mol\ H_2$ [7].

As the catalyst lifetime strongly depends on the organic metal content of the feed and the catalyst lifetime is an overriding item in reactor type selection between fixed bed and slurry bed. This aspect is treated first.

11.3.2 HDS reactor type selection

This HDS selection case is for crude oil fractions with such a low organic metal content that the catalyst lifetime is over 1 year.

All critical performance phenomena requirements for the reactor selection stated in Section 11.2 are then:
1. No hydrogen diffusion limitation inside the porous catalyst for long catalyst life
2. Liquid plug flow residence time distribution for deep conversion in limited reactor volume
3. Gas plug flow residence time distribution of the gas for effective adiabatic heat removal
4. Heat management of the exothermic reaction

The preliminary choice is then a fixed bed reactor operating in cocurrent gas-liquid flow. The particle size follows from an analysis by Satterfield [8]. The conclusion is that for particles in the order of 1 mm the effectiveness factor is 1. So, internal mass diffusion limitation is absent.

Satterfield furthermore concludes from mass transfer correlations that external mass transfer limitation for hydrogen from the gas phase to the liquid phase and from the liquid bulk phase to the catalyst particle surface are also not occurring [8].

Because the conversion of R-SH is sensitive to the local concentration of R-SH and of the hydrogen concentration, it means that the residence time distribution of liquid flow should be near plug flow.

The reaction heat can be removed from the reactor by applying a surplus of hydrogen in the reactor around a factor 10. The adiabatic temperature rise is then in the order of 10 °C. Section 25.3 provides more detailed design information for this case.

The gas phase residence time distribution is for this industrial design not critical because the hydrogen flow is in large surplus, so that the hydrogen concentration is nearly uniform in the reactor.

11.3.3 HDM reactor type selection

The reactor type to be selected for HDM with a catalyst time of a month or less is according to the method described in Section 11.2 a slurry reactor type. Shell, however, developed a moving bed trickle-flow reactor in which under reaction conditions spent catalyst is withdrawn from the bottom and fresh catalyst is added at the top. In reaction engineering terms, the reactor behaves as an adiabatic trickle-bed reactor in the same way as the HDS reactor. This reactor type was applied once at commercial scale with a start-up in 1989 [9]. Details of the reactor development and scale-up are provided by Harmsen [6].

All other commercial-scale HDM reactors are slurry type reactors with a heat exchanger inside the reactor. No details are provided on the reactor liquid residence time distribution, or the solids residence time distribution. The conversion range is 55 = 80. For a conversion of 55 %, a fully back-mixed reactor volume needed compared to a plug flow reactor for a first-order reaction is only 1.5 times a plug flow reactor. For a conversion of 80 % it requires, however, a factor 4 larger volume. Then having two reactors in series could make sense. Whether this is applied on a commercial scale is not reported.

This slurry reactor for HDM has been applied 8 times at commercial scale [10]. So, this reactor type can be considered a common practice and confirms the selection method of Section 11.2.

Bibliography

[1]   Dewes RM, Mc Carogher K, Van Olmen J, Kuhn S, Van Gerven T. Optimization of design and operational parameters of continuous oscillatory baffled reactors. Journal of Advanced Manufacturing and Processing. 2025 Jan;7(1):e10185.

[2]   van Swaaij WP, van der Ham AG, Kronberg AE. Evolution patterns and family relations in G–S reactors. Chemical Engineering Journal. 2002 Nov 28;90(1–2):25–45.

[3]   Taulamet MJ, Mariani NJ, Barreto GF, Martínez OM. Prediction of thermal behavior of trickle bed reactors: The effect of the pellet shape and size. Fuel. 2017 Aug;15(202):631–40.

[4] Stegeman D, van Rooijen FE, Kamperman AA, Weijer S, Westerterp KR. Residence time distribution in the liquid phase in a cocurrent gas–liquid trickle bed reactor. Industrial & Engineering Chemistry Research. 1996 Feb 8;35(2):378–85.

[5] Maiti RN, Nigam KD. Gas–liquid distributors for trickle-bed reactors: A review. Industrial & Engineering Chemistry Research. 2007 Sep 12;46(19):6164–82.

[6] Harmsen J, Bos R. Multiphase Reactors: Reaction Engineering Concepts, Selection, and Industrial Applications. De Gruyter; 2023.

[7] Jaffe SB. Kinetics of heat release in petroleum hydrogenation. Industrial & Engineering Chemistry Process Design and Development. 1974 Jan;13(1):34–9.

[8] Satterfield CN. Trickle-bed reactors. AIChE Journal. 1975 Mar;21(2):209–28.

[9] Oelderik JM, Sie ST, Bode D. Progress in the catalysis of the upgrading of petroleum residue: A review of 25 years of R&D on Shell's residue hydroconversion technology. Applied Catalysis. 1989 Feb 1;47(1):1–24.

[10] Arora A, Mukherjee U. Refinery configurations for maximising middle distillates. Petroleum Technology Quarterly. 2011;16(4):75–83.

12 Solids conversion reactors

12.1 Basic solids conversion reactors

Solids conversions take place in many commercial scale reactors such as in mineral ore processing [1], coal combustion [2], and catalyst regeneration [3]. This type of conversion is different from all other reactions treated so far in this book. The solid particles travel each separately through the reactor. In chemical reaction engineering, this is called segregated flow. It means that each particle carries its own reaction history. As a consequence, the residence time distribution (RTD) of the solids flow plays a very important role in reaching deep conversion. Even for zero-order reaction kinetics, the solids residence time distribution (RTD) affects the degree of conversion, as the reader can easily see by considering a fraction of particles with a very long residence time; with complete conversion reached far for this long residence time. While particles with a short residence time fraction have a low conversion. This low conversion is not compensated for by the particles with a very long residence time and complete conversion.

Also, the residence time distribution of the gas or liquid phase such as, for instance, oxygen in the gas phase reacting with coke on the catalyst particles to carbon dioxide. Often that reaction is not zero order in oxygen; hence, the gas phase RTD in general affects the degree of oxygen conversion.

For large scale solids conversion by a gas, such as in roasting mineral ores, where some of the reaction components are gaseous, mainly fluid bed reactors are applied.

Mass transfer from the fluid phase to the solid phase also often plays a role in the effectiveness of the reactor. If the intrinsic reaction rate is high, then it can be limited by mass transfer. That subject is treated in Chapter 10, so it is not treated in this chapter.

12.2 Reactor selection and design method

12.2.1 Reactor selection for solids conversion

Reactor selection for solids conversion starts with the selection between a fixed bed, moving bed, fluidized bed (gas solids), slurry (liquid-solid, or gas, liquid, solid), and between batch and continuous operation. For the latter choice, the rule of thumb of Douglas can be applied. It means that for a capacity above 10,000 t/y, continuous processes are to be chosen. For small capacity processes, such as in pharma and fine chemicals with a capacity below 5,000 t/y, batch process operation can be chosen [4].

For solids processing in large scale continuous operation, three reactor types are applied: slurry reactors, fluid bed reactors, and moving beds. The latter is, for instance, used in blast furnaces for iron production from iron-oxide minerals.

If deep conversion of (components on) the solids are needed such as in roasting process, then a narrow residence time distribution of the solids flow is required. As a

https://doi.org/10.1515/9783111203256-016

wide residence time distribution not only means that a very large, averaged residence time will be needed for conversion, but also the product quality will be less, as some particles will have a deep conversion and others a low conversion. So, there will be spread in the conversion of individual particles.

If the reaction is between the solids and a gaseous component, then a fluid bed reactor is the obvious choice. As a reasonably narrow RTD is required for deep conversion, a horizontal fluid bed divided in the chambers with small openings in the chamber walls for the solids to flow through is an attractive option.

If the reaction is between the solids and a liquid component, then a slurry reactor is the obvious choice.

Fluid bed reactors and slurry reactors in general have wide solids phase residence time distribution. Moreover, the solids flow, by nature, is a segregated flow system, which means conversion is strongly affected by the solids residence time distribution. Staging of the solids flow to narrow the RTD is therefore needed.

Staging of the solids flow can be obtained by having multiple reactors in series and by having a horizontal flow of the slurry, or fluidized medium and placing vertical panels with small holes for the slurry, or fluidization medium to pass true. The thus created chambers may be back-mixed, but a number of chambers in series create staged flow of the solids stream.

If the reaction requires a heat exchange, then a heat exchanger may be placed inside the reactor. For slurry reactors and fluid bed reactors, this is common practice.

12.2.2 Reactor selection for catalyst regeneration

If catalyst regeneration is simple, such as oxidizing coke on the catalyst by oxygen, then often the reaction will be carried out in the existing reactor for the main product. If, however, catalyst degrades rapidly then a separate reactor is chosen for catalyst regeneration. Figure 12.1 has been obtained from van Swaaij [5] that shows which reactor type is chosen in industry depending on the on-stream time of catalyst.

The best-known example for the last case is the Fluid Catalytic Cracking (FCC) process where coke is burned from the catalyst by oxygen in a separate fluid bed, called the combustor [6]. The majority of these combustor fluid beds are operated in the bubbling flow regime. The process was rapidly developed during the Second World War to produce gasoline from crude oil fractions. No consideration was given on the residence time distribution of the spent catalyst particles. The fluid bed combustor was just oversized with an average residence time of catalyst particles in the order of 1 hour, while the coke combustion of a single particle required a residence time of a few seconds.

Also, the residence time distribution of the gas phase in the fluid bed combustor appeared to be wide. Some of the gas moved as "bubbles" through the bed with a residence time of a few seconds, while some other parts of the gas moved through the bed via the

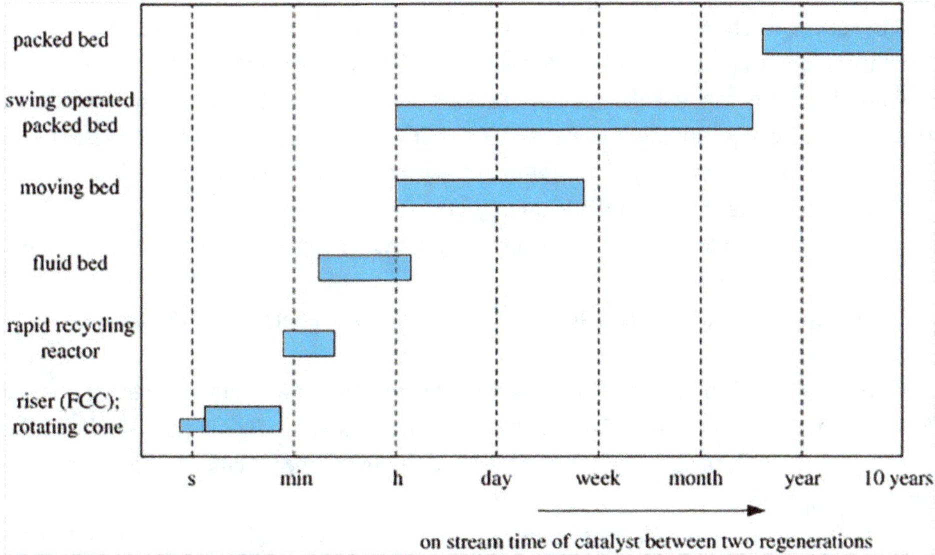

Figure 12.1: Reactor type applied versus on stream time catalyst.

"dense phase" with a residence time of several minutes. The negative effect of this wide gas phase RTD was compensated by a surplus of oxygen fed as air flow.

Recently, a riser fluid bed combustor for catalyst regeneration was selected [7]. This reactor type has a narrower solids residence time distribution and also a narrower gas phase RTD [8], but the solids flow is not in plug flow, as in the center solids flow upwards, while near the wall some of the solids flow downwards [9]. At the bottom of the riser, the solids volume fraction is higher, and the fluid bed behavior is still a dense phase flow, as the solids have to pick up the higher velocity [10].

The improved RTD of both phases, however, results in much smaller reactor volume and a deeper average coke conversion and a smaller spread in coke conversion of individual catalyst particles than in a bubbling fluid bed. Such a riser combustor is now in commercial scale operation in China [7].

12.3 Example case: Industrial oil shale conversion in horizontal fluid bed

This industrial case is about designing a process for oil shale conversion to crude oil and in particular about two solids conversion reactors of that process. In the 1980s of the last century, Shell wanted to have a process for the conversion of oil shale into crude oil. To that end, they chose Heinz Voetter, their most experienced and best chemical engineer, to design a concept and develop that concept for commercial scale production. Tarfaya oil shale of Morocco was chosen as a promising option, as it could be surface mined and the government of Morocco was interested in the project with Shell. After 2 years of studying, all literature on oil shale conversion and using his insight in chemical reaction engineering for commercial scale applications, he concluded that the cost of crude oil from oil shale could only be lower than all the other oil shale processes reported if:

A) The production capacity of the commercial process is much larger than the reported processes to benefit from the economy of scale.

B) The oil yield on kerogen (the hydrocarbon polymer inside the shale) is close to the maximum yield.

C) Resulting coke on the spent shale is used to generate energy for the process.

Conclusion A—let him design a fluid bed reactor concept for kerogen conversion. As the kerogen content of oil shale is 5–10 % (mass), it means that the whole process should be based on fluidization, as only then a very large single train capacity can be obtained at low investment cost.

Conclusion B—let him to a near plug flow horizontal shallow fluid bed reactor.Please review the previous sentence for clarity and accuracy. The plug flow ensured deep conversion of kerogen. The shallow bed ensured a short oil vapor residence time with little after cracking to undesired components [12, 13].

Conclusion C—let him to a fluid bed riser combustor reactor with again near plug flow behavior for deep coke conversion and also a narrow gas residence time distribution so that no large surplus of air was needed; so lower compression cost. Please review the previous sentence for accuracy. The spent hot shale was then used to heat the oil shale reactor and to preheat and dry the fresh shale [12, 13]. Transport of the solids from the combustor the other fluid beds was also under fluidized conditions.

The concept was worked out in a complete process design, and patents were drafted and obtained [11, 12]. After that, the project went into development. Main elements were kerogen conversion kinetics, coke combustion kinetics, heat transfer, mixing rates of hot spent solids in the kerogen fluid bed retort, and solids transport and control between the fluid beds. A process design package was made and then the project was mothballed, as crude oil prices dropped far below the processing and investment cost per ton of crude oil.

It is remarkable that the chemical engineer Heinz Voetter invented a superior process based on chemical engineering principles while the main ingredient fluid bed reactors were known for over 40 years in fluid catalytic cracking. Even plug flow shallow horizontal fluid beds were known at that time in the food industry for drying and roasting solid particles [13–16].

This whole story emphasizes the usefulness of applying chemical engineering design principles in the concept stage of a new process involving solids conversion.

Bibliography

[1] Zhang S, Zhou T, Li C, Zhang M, Yang H. Research progress and prospect of fluidized bed metallic ore roasting technology: A review. Fuel. 2024 Dec 15;378:132717.

[2] Scala F, editor. Fluidized Bed Technologies for Near-Zero Emission Combustion and Gasification. Amsterdam: Elsevier; 2013.

[3] Moulijn JA, Makkee M, Van Diepen AE. Chemical Process Technology. John Wiley & Sons; 2013.

[4] Douglas JM. Conceptual Design of Chemical Processes. New York: McGraw-Hill; 1988.

[5] van Swaaij WP, van der Ham AG, Kronberg AE. Evolution patterns and family relations in G–S reactors. Chemical Engineering Journal. 2002 Nov 28;90(1–2):25–45.

[6] de Vasconcelos PD, Mesquita LA. Gas-solid flow applications for powder handling in industrial furnaces operations. In: Ahsan A, editor. Heat Analysis and Thermodynamic Effects Heat Analysis and Thermodynamic Effects. Rijeka: Intech Open Access Publisher; 2011. Chapter 10.

[7] Bai D, Zhu JX, Jin Y, Yu Z. Simulation of FCC catalyst regeneration in a riser regenerator. Chemical Engineering Journal. 1998 Dec 2;71(2):97–109.

[8] Venderbosch R. The role of clusters in gas-solids reactors. PhD thesis. The Netherlands: U. Twente; 1998.

[9] Smolders K, Baeyens J. Overall solids movement and solids residence time distribution in a CFB-riser. Chemical Engineering Science. 2000 Oct 1;55(19):4101–16.

[10] Issangya AS, Karri SR, Cocco RA, Knowlton T, Chew JW. Solids flux profiles of Geldart Group A particles in high-velocity circulating fluidized bed risers. The Canadian Journal of Chemical Engineering. 2023 Jan;101(1):256–68.

[11] Harmsen J, Bos R. Multiphase Reactors: Reaction Engineering Concepts, Selection, and Industrial Applications. De Gruyter; 2023.

[12] Voetter H, Van Meurs HC, Darton RC, Krishna R. Process for the extraction of hydrocarbons from a hydrocarbon-bearing substrate and an apparatus therefore. United States patent US 4,439,306. 1984.

[13] Voetter H, Darton RC, Van Meurs HC, Krishna R. Process for the combustion of coke present on solid particles and for the production of recoverable heat from hydrocarbon-bearing solid particles and apparatus therefor. United States patent US 4,508,041. 1985.

[14] Hua L, Zhao H, Li J, Zhu Q, Wang J. Solid residence time distribution in a cross-flow dense fluidized bed with baffles. Chemical Engineering Science. 2019 Jun 8;200:320–35.

[15] Mujumdar AS, Devahastin S. Applications for fluidized bed drying. In: Yang WC, editor. Handbook of fluidization and fluid particle systems. New York: Marcel Dekker; 2003. pp. 469–84.

[16] Khanali M, Giglou AK, Rafiee S. Model development for shelled corn drying in a plug flow fluidized bed dryer. Engineering in Agriculture, Environment, and Food. 2018 Jan 1;11(1):1-8.

13 Polyolefin polymerization reactors

13.1 Basics polyolefin polymerization reactors

In polyolefin polymerization, olefin monomers react to polymer chains. Most common olefins used are ethylene, propylene, styrene, and butadiene [1].

The chain propagation reaction is catalyzed by a solid catalyst. The chain propagation is terminated by a catalyzed reaction with hydrogen. The ratio of hydrogen to olefin concentrations at the catalyst and the temperature determine the chain length. Chain length and chain length distribution are very important properties for polymer applications. This means that the olefin, hydrogen concentration ratio, and the temperature at each catalyst site has to be uniform inside the whole reactor and that it must controlled [1].

The gas phase olefin and the hydrogen molecules have to be diffused through the polymer particles to the catalyst sites; hence, mass transfer inside the polymer particles also plays a role [1].

The propagation reaction is highly exothermic [1]. This means that heat must be removed from the reactor. It also means that the catalyst particle and its surrounding polymer should not be overheated. This implies that the reaction rate and the polymer particle size must be designed such that this overheating does not occur.

So, the reactor designer has to ensure that the reactor conditions that at the polymer particle scale and at the total reactor scale are uniform for monomer concentrations and also the temperature at the particle scale and the total reactor scale must be uniform [2].

The desired polymer production capacity is in general in the order of hundreds of kton/year [1]. Hence, a reliable design of the commercial scale reactor system is important.

As polymer grades are developed all the time for new applications, it means that a well and down-scaled pilot plant for testing these new polymer grades is also needed. Also, electrostatic countermeasures have to be tested at a pilot plant. Hence, a well scaled-down pilot plant is needed.

13.2 Polyolefin reactor design method

13.2.1 Design base kinetics

There are two main reactions involved in polymerization in general; chain propagation and chain ending. A catalyst radical form a monomer radical. The radical reacts with a monomer molecule M and forms a polyolefin chain CH. This reaction keeps going and is called chain propagation.

https://doi.org/10.1515/9783111203256-017

The reaction scheme reads as

$$CH + M = CH\text{-}M$$

This chain propagation is repeated until it ends when the catalyst leaves the molecule and starts a different chain. This phenomenon is called chain transfer and affects molecular weight distribution stochastically. Hydrogen terminates the propagation by the reaction with stoichiometry,

$$[CH]_n + H_2 = CH_{n+2}$$

The average chain length is then determined by the chance of meeting a monomer or a hydrogen molecule. The average chain length is therefore determined by the concentration ratio of monomer/H_2. Meier confirmed this experimentally [3].

The propagation reaction is given as first order in monomer, and catalyst concentration (or active radical sites concentration. The chain propagation rate expression is

$$r_p = k_p M \, Cat^* \tag{13.1}$$

M is the monomer concentration. Cat^* is the sum of the active catalyst and radical site at the end of the polymer chain.

The reaction rate expression follows first-order kinetics in hydrogen and activated chain end [1]:

$$r_e = k_e H_2 C^* \tag{13.2}$$

The ratio of r_p/r_e determines among other effects the chain propagation over the chain termination; hence, affects the molecular weight of the polymers [1]. This ratio shows that the molecular weight can be controlled by the monomer over the hydrogen concentration in the absence of mass transfer limitation of the monomers inside the polymer particle.

The catalyst decays in time. The decay can be described as a first-order decay rate [1]. But more complex decay kinetics are also available from Soares [1]. Meier observed experimentally that the presence of hydrogen reduces the decay rate [3]. For our case study with a focus on the design choice for the solids phase residence-time distribution. Please review the previous sentence for accuracy and clarity.These additional effects are not important. So, we use the first-order decay kinetic equation,

$$r_d = -k_d C^* \tag{13.3}$$

The effects of temperature on the three reactions are provided by the Arrhenius equation:

$$k_p = k_{p0} e_p^{-E}/RT \tag{13.4}$$

$$k_e = k_{e0}e_e^{-E}/RT \qquad (13.5)$$

$$k_d = k_{d0}e_d^{-E}/RT \qquad (13.6)$$

The parameter values are determined experimentally. However, the value of k_0 will be very uncertain as this is the reaction rate constant k at T = infinite. So, this is far outside the experimental range. It is therefore far better to use a k_{ref} value at a reference temperature within the experimental range and rewrite the expression as

$$\ln(k/k_{ref}) = (E/R)(1/T_{ref} - 1/T) \qquad (13.7)$$

13.2.2 Reactor type selection

We use the Douglas rule of thumb that for process capacities >10,000 t/a by which a continuously operated reactor type is to be chosen [4]. As for polymer processes, the capacity is in general much larger than 10,000 ton/y; this means continuous processing is to be chosen.

Furthermore, as large amounts of solids have to be processed, a fluid bed reactor is the obvious choice.

13.2.3 Concept reactor design

The critical performance phenomena to be taken into account for reactor selection are gas-phase residence time distribution, solids (catalyst and polymer) residence time distribution, olefin mass transfer to catalyst, temperature control of particles and reactors.

13.2.3.1 Gas phase residence time distribution and reactor design

The concentrations of hydrogen and monomer should be uniform throughout the reactor to get a uniform polymer chain length. The most reliable way of achieving this is to have a reactor with a large gas recycle to feed ratio. With a recycle ratio of 100, the maximum concentration difference between reactor inlet and outlet is 1 %. So, this recycle ratio is chosen.

The additional advantage of this large recycle ratio is that the gas phase residence time distribution is fully back mixed, so no complex gas phase model is needed for the reactor sizing design.

This gas phase external recycle is easily obtained by a cyclone separator, followed by a gas compressor.

13.2.3.2 Solids phase residence time distribution and reactor

The solids flow residence time distribution is not relevant for the polymer chain length distribution, if the gas phase composition is uniform throughout the reactor, as treated in Section 13.2.3.1. The solids flow RTD is, however, relevant for the catalyst needed per ton of product and for the particle size distribution. So, some staging to reduce catalyst cost per ton of product is desired.

This staging of solids flow is treated in Chapter 11. It provides options for solids staging in fluid beds, such as a horizontal crossflow fluid bed. It may have vertical baffles for additional prevention of back-mixing behavior.

13.3 Example case: Concept reactor design polyethylene

13.3.1 Reaction kinetics

There are two main reactions involved: chain propagation and chain ending. We take polyethylene formation for the description. Other polyolefins have analogue chemistry. A catalyst radical forms an ethylene radical. The radical reacts with an ethylene molecule and forms a polyethylene. This keeps going and is called chain propagation.

The reaction scheme and stoichiometry of chain propagation:

$$C_2H_4 + C_2H_4 = [C_2H_4]_2$$

This chain propagation is repeated until it ends when the catalyst leaves the molecule and starts a different chain. This phenomenon is called chain transfer and affects molecular weight distribution stochastically. Hydrogen terminates the propagation by the reaction with stoichiometry,

$$[C_2H_4]_n + H_2 = C_{2n}H_{4n+2}$$

The average chain length is then determined by the chance of meeting a monomer or a hydrogen molecule the average chain length is therefore determined by the concentration ratio of monomer/H_2. Meier confirmed this experimentally [3].

The propagation reaction is given as first order in monomer, and catalyst concentration (or active radical sites concentration). The chain propagation rate expression is

$$r_p = k_p M \, Cat^* \tag{13.8}$$

M is the monomer concentration. Cat^* is the sum of the active catalyst and radical site at the end of the polymer chain.

The reaction rate expression follows first-order kinetics in hydrogen and activated chain end [1]:

$$r_e = k_e H_2 C^*$$ (13.9)

The ratio of r_p/r_e determines among other effects the chain propagation over the chain termination, and hence affects the molecular weight of the polymers [1]. This ratio shows that the molecular weight can be controlled by the monomer over the hydrogen concentration in the absence of mass transfer limitation of the monomers inside the polymer particle.

The catalyst decays in time. The decay can be described as a first-order decay rate [1]. But more complex decay kinetics are also available from Soares [1]. Meier observed experimentally that the presence of hydrogen reduces the decay rate [3]. For our case study with a focus on the design choice for the solids phase residence-time distribution, these additional effects are not important. So, we use the first-order decay kinetic equation,

$$r_d = -k_d C^*$$ (13.10)

The effects of temperature on the three reactions are provided by the Arrhenius equation:

$$k_p = k_{p0} e_p^{-E}/RT$$ (13.11)

$$k_e = k_{e0} e_e^{-E}/RT$$ (13.12)

$$k_d = k_{d0} e_d^{-E}/RT$$ (13.13)

The parameter values are determined experimentally. Soares studied experimentally polyethylene formation kinetics [1]. The precise analysis of the experimental results to obtain the kinetic parameter values is however not clear, so his deactivation parameter values cannot be used reliably either. In general, analyzing experimental results to obtain k_0 and E values is not a reliable way of obtaining parameter values.

The value of k_0 will be very uncertain as this is the reaction rate constant k at T = infinite. So, this is far outside the experimental range. It is therefore far better to use a k_{ref} value at a reference temperature within the experimental range and rewrite the expression as

$$\ln(k/k_{ref}) = (E/R)(1/T_{ref} - 1/T)$$ (13.14)

So, the parameter values of Soares cannot be used reliably. We therefore provide his experimental parameter values at one temperature, to be used for a case illustration and exercise in Table 13.1.

Table 13.1: Kinetic parameter values of polyethylene formation [1].

Parameter	Dimension	Value	Comments
E_p	$cal\,mol^{-1}$	$1.4\,10^4$	
E_d	$cal\,mol^{-1}$	$1.5\,10^4$	
R	$cal\,mol^{-1}\,K^{-1}$	1.98717	Universal gas constant
$k_p\,M$	s^{-1}	$1.13\,10^5$	$T = 393\ P = 20\,10^5\ N/m^2$
k_p	$s^{-1}\,m^3/mole$	160	
k_d	s^{-1}	$1.4\,10^{-4}$	$T = 393$
M	$mole/m^3$	700	$M = n/V = P/RT = 20\,10^5/(8\,.\,363) = 700$
C^* (fresh)	mol/m^3	10^{-5}	This is per unit reactor volume; not particle volume
$k_p\,C^*$	s^{-1}	$160\,10^{-7}$	This is per unit reactor volume; not particle volume

13.3.2 Reactor design

The competition in polyethylene market is global and fierce and the catalyst costs are a significant part of the production cost. Management therefore decides to ask the chemical reaction engineer to generate a new reactor with a lower catalyst cost. The other requirement is a commercial scale reactor design for 100 kt/y. This means a polyethylene production of 120 mol/s with the assumption of 8,000 h production time per year.

The chemical reaction engineer concludes that the present reactors are back mixed regarding the catalyst particles, which means that part of the catalyst quickly leaves the reactor, while other catalyst particles stay very long in the reactor. This means that more catalyst per ton of product is needed than when the catalyst would flow in near plug flow.

The chemical reactor engineer therefore chooses a novel reactor type, namely a crossflow horizontal column fluid bed with vertical baffles for solids flow staging. This results in a staged solids residence time distribution requiring less catalyst per ton of product as the catalyst travels as staged flow through the reactor. Fresh catalyst is fed at the beginning and spent catalyst is withdrawn with the product at the reactor outlet.

The gas flow leaving the reactor via a cyclone is cooled by an external heat exchanger and recycled back to the reactor. The gas recycle flow rate is >20 to the ethylene feed flow rate, which means that the concentration of hydrogen inside the reactor is nearly uniform. So, the polymer chain length is nearly uniform. This part of the design is the same as present commercial scale fluid bed reactors. So, the temperature in the reactor is nearly uniform by the large gas recycling with the external heat exchange cooler.

The reactor volume required is calculated the information of Table 11.1. The reaction rate for ethylene is 1.0 mol/(m^{3}s). The required product is 120 mol/s so the reactor volume required is 120 m^3.

Because the reactor operates at 20 bar, the reactor diameter will be limited to 4 m. The required reactor length is then 13 m. Staging of the solids flow is then easily feasible using vertical baffles. The reactor is then similar to a crossflow horizontal bubble column reactor.

A fall back-design is a conventional fluid bed reactor for polyethylene described by Soares [1] and others [5]. The advantage of the solids plug flow behavior for a lower catalyst demand has to be evaluated with the development cost for the new reactor type.

The catalyst feed rate to achieve the desired catalyst concentration of 10^{-5} mol/m^3 is not determined for this concept design. It needs details of the catalyst particle active site concentration, which are not readily available.

The gas recycling needed for sufficient heat removal is also left out. A good review article by Alves [5] shows that it is feasible and that a near uniform temperature inside the reactor is obtainable.

Bibliography

[1] Soares JB, McKenna TF. Polyolefin Engineering. Chicester, UK: Wiley-VCH; 2012.
[2] Harmsen J, Bos R. Multiphase Reactors: Reaction Engineering Concepts, Selection, and Industrial Applications. Berlin: De Gruyter; 2023.
[3] Meier GB. Fluidized bed reactor for catalytic olefin polymerization. PhD thesis. Enschede: University of Twente; 2000.
[4] Douglas JM. Conceptual Design of Chemical Processes. New York: McGraw-Hill; 1988.
[5] Alves RF, et al. Gas-phase polyethylene reactors—A critical review of modeling approaches. Macromolecular Reaction Engineering. 2021;15(3):2000059.

14 Reactor heat management design

14.1 Basics reactor heat management

Chemical reactions in general produce heat or consume heat. Heat producing reactions are called exothermic and heat consuming reactions are called endothermic. If the reactor does not exchange heat by a heat exchanger, then it is called an adiabatic reactor. If it exchanges heat via a heat exchanger, then it is called a heat exchanger reactor, or a cooled reactor. It may then also be called "an isothermal reactor," as the reactor temperature is nearly uniform. The heat exchange effect, however, takes place by a temperature difference between the reactor medium and the heat exchanger, so there is a small temperature gradient inside the reactor. The reaction heat can also be cooled by evaporation of a component inside the reactor. Sometimes the resulting vapor is cooled in condenser, and the cool liquid is returned to the reactor, so that an additional cooling effect is obtained.

The heat produced or consumed is defined by the reaction enthalpy ΔH_r. For exothermic reactions, the reaction enthalpy has a negative value. For endothermic reactions, it has a positive value.

Reaction enthalpy does not have fixed value, but it depends on the temperature at which the reaction takes place. The reaction enthalpy in databases is therefore defined at a fixed temperature, often at 298 K. This means that the reaction enthalpy has to be determined experimentally for the actual reaction conditions. This holds especially for gas phase reactions, where the heat capacities are sensitive to temperature [1].

Nowadays flow sheet programs have physical property models, which take into account the temperature effects on the reaction enthalpy. However, if the reaction is highly exothermic, or endothermic and the reaction heat is used via heat integration with other process steps, then it is important to validate the reaction enthalpy experimentally.

Reaction heat management is needed in the concept design stage for several reasons. The first reason has to do with reactor safety. Reaction rates in general increase with temperature. If the reaction is exothermic, the reactor temperature will in general increase when the reaction progresses. Beyond a certain temperature, the reaction rate increases so fast that its control is no longer feasible. The reaction then runs away to a very high temperature. It may cause a reactor meltdown, or a reactor explosion if the pressure also increases with increasing temperature. This subject is addressed in Section 14.2.

The second reason is that the temperature window favoring the desired reaction over the undesired reactions is in general narrow.

The third reason is that often the reaction is an equilibrium reaction and the reaction equilibrium constant depends on the temperature. Hence, temperature control is needed to achieve a desired equilibrium value.

For these reasons, heat management by design is needed so that the temperature stays in the desired temperature window. If this heat management design is such that

https://doi.org/10.1515/9783111203256-018

the reactor temperature is nearly uniform over the whole reactor, the reactor is called isothermal.

14.2 Reactor heat management design method

14.2.1 Reaction temperature and composition design to prevent runaway

Evaluating a reactor concept design on safety is in general focused on evaluating the runway reaction and connected explosion risk. In an exothermic reaction, the heat produced increases the temperature. This increases the reaction rate and thereby the reaction heat production rate. Beyond a certain temperature, this heating up goes so fast that the temperature can no longer be controlled and the reaction runs away. This is known as "thermal runaway." It can cause an explosion. Common triggers for such an explosion in existing reactors are a new feed ingredient, a new catalyst, or a new solvent, or operating outside the design window. For these reasons, runaway reaction experiments in dedicated laboratory equipment are essential. It is beyond the scope of this book to treat this subject in detail.

14.2.2 Selection between an adiabatic or a cooled reactor

Adiabatic reactor temperature rise calculation

Heat exchangers are in general expensive. Therefore, first the option of an adiabatic reactor is evaluated. To that end, the adiabatic reaction temperature rise is calculated. This is a simple heat balance over the inlet and the outlet streams, as there is no heat exchange with the environment.

The adiabatic reaction temperature rise (or drop) is given by Levenspiel [1], shown by equation (14.1):

$$\Delta T_{ad} = -\Delta H_r X / [Cap^{out} + (Cap^{out} - Cap^{in})] \tag{14.1}$$

with
ΔH_r: The reaction enthalpy per mole of entering reactant A (J/mol A)
X_A: the degree of conversion of entering reactant A
Cap^{out}: mean specific heat of the total fluids leaving the reactor/mol of entering reactant A
Cap^{in}: mean specific heat content of the total fluids entering the reactor/mol of entering reactant A

With expression (8.1), the adiabatic temperature rise can now be estimated.

Table 14.1: Specific heat capacity of hydrogen and hydrocarbon gases.

Specific heat capacity C_p	Value in kJ/(kg °K) at 298 °K
Hydrogen gas	14
Hydrocarbon gas	2
Hydrocarbon liquid	2

For a quick estimate of the adiabatic temperature rise Table 14.1 of specific heat capacity coefficients C_p for hydrogen gas and hydrocarbon gas and liquids may be helpful.

If the calculated adiabatic temperature rise is too high to select an adiabatic reactor design, then two options are available to reduce the adiabatic temperature rise.

The first option is to reduce the value of conversion X. This, however, means that unconverted feed component A from the outlet stream has to be separated from the stream and recycled back to the reactor inlet. So, the investment cost and energy required increase for this separation. If the separation means just enlargement of the separation of A from the product stream, then this option can be pursued. If it needs an additional separation step, then this option in general is not attractive.

The second option is to add (more of) another fluid such as a solvent of a gas, or fine solids (in fluid bed) reactor to the feed, so that the cap value increases. Again, this means that the other fluid has to be separated from the reactor outlet stream. If separation is already needed and it only means an enlargement of that separation duty, then this option can be chosen. This option is, for instance, chosen for hydrogenation in oil refineries and bulk chemical processes. Several aspects favor this design application. It is the high specific heat capacity of hydrogen, the high reactor pressure, so that the mass flow of hydrogen is large, and third, the hydrogen is easily separated from the liquid product stream by a flash separator. The only expensive equipment is the hydrogen recycle compressor.

Applying this surplus of reactant is not feasible, then a reactor with heat exchange will have to be selected.

14.2.3 Reactor design with heat exchange

The adiabatic heat production rate given in expression (8.1) is a function of the temperature and of the reactant concentrations [2]. For a batch reactor or a plug flow reactor, the Heat Production Rate (HPR) of the reaction can be calculated as a function of the adiabatic temperature rise as the reaction progresses. It is plotted in Figure 14.1. In the first part of curve, the heat production rate increases exponentially because the reaction rate increases exponentially with temperate. In the last part of the curve, the heat production rate flattens, as the conversion of reactant nears completion.

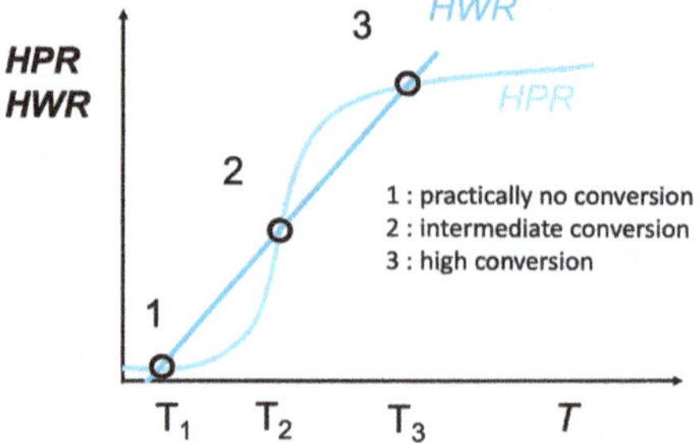

Figure 14.1: Heat production rate and heat withdrawal rate versus temperature. All points 1, 2, 3 are steady states. Point 2 is unstable [2].

The Heat Withdrawal Rate (HWR) by the heat exchanger is given by expression (14.2),

$$HWR = \alpha \,.\, \Delta T \,.\, HTA, \tag{14.2}$$

in which

α: the overall heat transfer coefficient $W/(m^2K)$
ΔT: The overall temperature difference between reactor and coolant K
HTA: The heat transfer area m^2

Figure 14.1 shows three points where the 2 lines have the same value for HPR and HWR. So, at those points, the heat production rate equals the heat withdrawal rate. We now investigate the operational stability of these points.

If at point 1 the temperature rises, the HWR will then be above the HPR line, so then the cooling rate is higher than the heat production rate, so the reactor temperature will drop back to the steady-state point 1. So, this is a stable operating point.

If at point 2 the temperature rises a little, the HWR line will be below the HPR line so the reaction temperature will further increase. This is called reaction runaway. So, point 2 is an unstable operating point.

At point 3, the HWR is above the HPR line, just as at point 1, so this is also a stable operating point.

The HWR line can be changed by the choice of the cooling temperature and by the choice of the heat transfer area. The HPR can be changed by the choice of inlet temperature and the use of diluents in the feed to the reactor. Figure 14.1 can be the starting point of a cooled reactor design.

The safe sizing of a reactor to avoid running away is however beyond the scope of this book. That not only involves a steady-state solution design but also involves a sensitivity analysis to uncertainties in the physical and thermodynamic properties, design parameters, and reactor model, but it also involves experimental testing of the runway behavior of the chemical reaction compositions.

14.2.4 Sizing catalytic particles to avoid a high internal temperature

For exothermic heterogeneously catalyzed reactions, the temperature inside the catalyst particle may be higher than the surrounding fluid flowing around it.

This can be calculated by an expression from Levenspiel [1]. This expression contains the concentration of component A in the center, which is a complex function of the Thiele modulus. For simplicity, we provide here a conservative approach for cases with Thiele modulus $\phi < 0.5$. The concentration difference of A between the particle skin and the center is then <0.1 (see Figure 18.4, p. 384, Levenspiel [1]).

A conservative estimation of the temperature difference between the particle center and the particle skin for $\phi < 0.5$ is then given by expression (14.3),

$$\Delta T_{particle} = D_e 0.1\, C_{Askin}(-\Delta H_{rA})/k_{eff}, \tag{14.3}$$

in which

D_e: The effective diffusion coefficient of component A inside particle (m²/s)
C_{Askin}: The concentration of A at the particle skin (mole/m₃)
$-\Delta H_{rA}$: The reaction enthalpy of component A (J/mole)
k_{eff}: The effective particle heat conductivity (W/(m°K))

If this temperature difference is higher than 1 °K, then the reaction temperature should be reduced such that the temperature difference will be negligible.

The temperature in the bulk of the fluid can be higher than the particle skin temperature. This can be estimated by expression (14.4),

$$\Delta T_{particle} = L r_A(-\Delta H_{rA})/h, \tag{14.4}$$

in which

$\Delta T_{particle}$: temperature difference between bulk fluid temperature and particle skin (°K)
L: characteristic particle length. For spheres: $L = d_{particle}/6$ (m)
h: heat transfer coefficient for fluid flow around particles (W/(m²K))

Chapter 22 contains expressions for calculating heat transfer coefficients around particles.

If the temperature difference is higher than 1 °K, then the reaction temperature can be reduced to such a level that the temperature difference becomes less than 1 °K. Alternatively. the fluid flow velocity may be increased, or the catalyst particle size reduced so that this temperature difference becomes negligible.

The reasoning for designing with no temperature differences between the bulk fluid and the catalyst particle skin and particle center is that for a higher catalyst temperature than the fluid flowing around the reaction selectivity for the desired product will in general be lower. This holds if the by-product reactions increase more by the higher temperature than the main reaction, which is often the case.

14.3 Example case: Ethylene glycol from ethylene oxide and water

14.3.1 Ethylene glycol reaction data

Ethylene oxide (EO) reacts with water to form mono ethylene glycol (MEG).

MEG is an important industrial chemical and also reaction kinetics and reaction runaway behavior have been studied and reported by Melhem et al. [3]. Here are the main results of Melhem's study. The reaction stoichiometry is

$$EO + H_2O = MEG$$

The reaction is exothermic with a reaction enthalpy; $\Delta H_r = -97\,kJ/mol$.

There are undesired consecutive reactions to Di-Ethylene Glycol (DEG) and Tri-Ethylene Glycol (TEG) with the following reaction stoichiometry:

$$MEG + EO = DEG$$

$$DEG + EO = TEG$$

The reaction enthalpy of these consecutive reactions is slightly higher than for the main reaction.

The reaction kinetics of the desired reaction is first order in EO and water. All glycol components, lumped as ROH, have a second-order effect:

$$R_{EO} = k\,EO\,H_2O\,ROH^2 \tag{14.5}$$

$$k = k_0\,\exp(-E_a/RT) \tag{14.6}$$

The consecutive reactions have similar kinetics to the first reaction. Mono Ethylene glycol and Di-ethylene glycol react with EO in the same way as EO reacts with water [Melhem].

14.3.2 Reactor type selection

To get a high selectivity of the desired product MEG, a large surplus of water is desired, as the first reaction is first order in water concentration, while it has no effect on the consecutive reactions.

Because of undesired reaction is a consecutive reaction and first order in the MEG concentration, a plug flow reactor is desired over a back-mixed (CISTR) reactor. The latter the high outlet MEG concentration everywhere, while the plug flow reactor starts with a zero concentration of MEG, which increases until it reaches the outlet concentration.

As the reaction is a homogeneous liquid phase reaction and the liquid is mainly water with a low viscosity, a pipe reactor operating in the turbulent flow regime is an obvious choice.

14.3.3 Heat management reactor design

To get a feeling of the problem, let us first calculate the adiabatic temperature rise for a large surplus of water over EO. As this surplus of water is preferred anyway for a selectivity reason, see Section 14.2.2. For simplicity, we take a 10 % (kg/kg) of EO in water as the feed composition. The molar concentration of EO at the inlet is then 2 mol/kg. The heat capacity of the feed stream is then 3.8 kJ/ kg feed mixture (0.1 × 2 kJ/kg EO + 0.9 × 4 kJ/kg H_2O).

The specific heat of the feed mixture parameter CAP is then 1.9 kJ/mol EO.

We neglected the second term in expression, as the heat capacity of the outlet stream will be similar to the inlet stream due to the large surplus of water. The adiabatic temperature rise is then 25 °C (97/3.8).

This is lower than the rule of thumb maximum for stable adiabatic operation of 30 °C. Hence, an adiabatic reactor is feasible, and the selected feed composition is 10 % EO in the feed mixture.

The inlet temperature should be chosen such that by-product formation is minimized.

14.3.4 Development program

An experimental runaway experimental program is needed to ensure that no other exothermic reactions occur. Such a program is also needed to define the optimum feed composition and inlet temperature and also to ensure that a dynamic reactor model can be made to define the reactor temperature control system. The results of an experimental program of the EO hydration to ethylene glycol is provided by Melhem [3].

Bibliography

[1] Levenspiel O. Chemical Reaction Engineering. New York: John Wiley & Sons; 1998.
[2] Harmsen J, Bos R. Multiphase Reactors: Reaction Engineering Concepts, Selection, and Industrial Applications. Berlin: De Gruyter; 2023.
[3] Melhem GA, Gianetto A, Levin ME, Fisher HG, Chippett S, Singh SK, Chipman PI. Kinetics of the reactions of ethylene oxide with water and ethylene glycols. Process Safety Progress. 2001 Dec;20(4):231–46.

Part IV: **Unit operations design**

15 Unit operations design

15.1 Basics unit operation design

Unit operations are the building blocks of process designs. After the functional process design step, unit operations are selected for the defined functions as described in Chapter 1. Design methods for unit operations are available from textbooks such as Perry's Chemical Engineers' Handbook [1]. In this book, we focus on shortcut design methods for quickly generating a feasible and attractive concept for most common unit operations in Chapters 16–23.

For each unit operation, the following basic elements critical to their performance are addressed:

- Thermodynamic equilibria and properties
- Mass transfer of components from one phase to another phase
- Residence time distribution of fluids
- Heat effects
- Heat transfer
- Impulse transfer

Ad thermodynamic equilibria and physical transport properties

Thermodynamic equilibria play an important role in separations such as distillation, extraction, absorption, and membrane separations. Thermodynamic equilibrium properties can be obtained from databases. If data are not available in databases, then they can be generated from physical property models of flow sheet packages, but the generated data should be validated by experiments.

Physical transport properties also play an important role in all unit operations designs. Values for them can in general be obtained from reference books and physical properties databases.

Ad mass transfer

Mass transfer between phases plays an important role in all separations. The rate of mass transfer determines the required size of the unit operation to reach the design goals. Dimensionless correlations for mass transfer are available for those unit operations.

Ad residence time distribution

Residence time distribution (RTD) is often not considered for unit operations other than reactors, but it is an important phenomenon for the performance of unit operations. Hence, RTD information should always searched for and if not available, should be determined experimentally.

https://doi.org/10.1515/9783111203256-020

Ad heat effects

Heats of evaporation, condensation, absorption, or adsorption can play a role in unit operations. In some cases, the effects are small and after an initial consideration, it can be neglected in further concept design.

Ad heat transfer

Heat transfer plays a role if heat effects play a role and when there are temperature differences inside the unit operation. Heat transfer is therefore nearly always a phenomenon to be taken into account in unit operation design.

Ad impulse transfer

Impulse transfer of the flowing fluids from one phase to the other and to internals should always be considered. For countercurrent flow, impulse transfer determines the velocity limits for the flows. The limits are called flooding.

Impulse transfer of the fluids on packings and internals should also be considered to avoid damage to the packings and internals on which the packing rests.

15.2 Unit operations design

15.2.1 Unit operations design introduction

The first step in unit operations design is that the design is for the envisaged commercial scale and not for the pilot plant scale. If the design is to be made for the pilot plant scale, then also a design for the commercial scale should be made, so that the pilot plant design is representative of the envisaged commercial scale process. Details of this scale-up method are found in Chapters 24 and 25.

Design of unit operations most used in industry are treated in Chapters 16–23. However, other unit operations may be needed for an integral process design. And also new unit operations are developed by research, notable process intensified unit operations [2]. When for one of these unit operations a design is to be made, then all five phenomena of Section 15.1 should be treated.

If information is lacking on one or more phenomenon, then a research program should be started or alternative unit operations should be considered for which all information is available.

15.2.2 Thermodynamic equilibria

Thermodynamic equilibria are important for designing separation unit operations, such as distillation, extraction, absorption, membrane separations. Thermodynamic equilibrium data can be obtained from databases such as the Dortmund database [2]. If data are not available in databases, then they can be generated from physical property models of flow sheet packages, but the generated data should be validated by experiments.

15.2.3 Mass transfer

Mass transfer between phases plays an important role in all separations. The rate of mass transfer determines the required size of the unit operation to reach the design goals. For most unit operations, dimensionless correlations for mass transfer rates are available.

If these correlations are not available, then still a first mass transfer rate can be determined by the Sherwood number Sh defined as

$$Sh = k\,d/D \tag{15.1}$$

k: (m/s) Mass transfer coefficient
d: (m) Characteristic diameter or length scale
D: (m^2/s) Diffusion coefficient component to be transferred

And the specific surface area a through which the mass transfer is occurring.

The minimum value for the Sherwood number is: Sh = 2.

The specific surface area a may be estimated from the characteristic length scale of the interface through the mass transfer takes place: For bubbles with diameter d_b: $a = 6/d_b$.

For film layers: with a layer thickness l: $a = 1/l$.

The mass transfer performance parameter $k_l a$ can now be estimated.

If mass transfer is for a transient case, then the mass transfer can be estimated with the Fourier number. The design method for these cases is found in Chapter 23.

15.2.4 Residence time distribution

Residence time distribution is often not considered for unit operations other than reactors, but it is an important phenomenon to be taken into account for all unit operation designs. If no information is provided for the unit operation, then an estimate often can be made by estimating the value of the Peclet number Pe and the Bodenstein number Bo:

$$Bo = vD/D_{disp} \tag{15.2}$$

$$Pe = L/D \tag{15.3}$$

v: (m/s) Linear averaged fluid velocity.
D: (m) Diameter of the object through which, around which, the fluid is flowing.
D_{disp}:   (m^2) Dispersion coefficient
L: (m) Length of unit operation

The Bodenstein number is a function of the Reynolds number. For turbulent flows, hence for flow around objects such as spheres turbulent flow is obtained for Re >100. (*Check. For flow through pipes and channels, turbulent flow is obtained for* Re > 2,000.)

For turbulent flow: Bo = 2.

So, now with (15.3) the value of Pe is known.

In other cases, such as multichamber mechanically stirred columns, the number of back-mixed stages can be equal to the number of chambers, provided that the channels for flow from one chamber to the next is narrow and the stirrer blades are aligned. The latter is needed to avoid pressure difference waves over the channels causing short-cutting of flows.

15.2.5 Heat effects

Heat effects play a role by evaporation, condensation, absorption, or adsorption. In some cases, the effects are small, and after initial consideration, can be neglected in further concept design.

Strangely enough, heat effects are often not treated unit operations such as adsorption but should always be considered, as absorption, adsorption, dissolution, and evaporation always have a heat effect, and thereby often change the local temperature. This in turn affects the equilibrium values, and hence affect the unit operation performance.

15.2.6 Heat transfer

Heat transfer plays a role if heat effects are present, and/or when there are temperature differences between the fluids entering the unit operation.

Heat transfer is in general described by a Nusselt number (Nu) as a function of the Reynolds (Re) and Prandtl number (Pr). Chapter 22 shows a correlation for heat transfer with these numbers.

If no correlation is available, then a conservative estimate can be made for the heat transfer coefficient α by the minimum value for Nu = 2.

15.2.7 Impulse transfer

Impulse transfer of the flowing fluids from one phase to the other and on internals should always be considered. For countercurrent flow, impulse transfer determines the velocity limits for the flows. The limits are called flooding. For cocurrent flow flooding does not occur, but impulse transfer determines the flow regime.

Impulse transfer of the fluids on packings should also be considered to avoid damage to the packings and internals on which the packing rests.

15.2.8 Limitations of design methods

Limitations to the presented design method is that it is only applicable to unit operations where all information on critical performance phenomena of Sections 15.2.2–15.2.7 are available.

If the information is not available, then conservative estimates can be made for some values. If the unit operation looks very attractive, then an experimental program can be started to determine phenomena and their relations with dimensionless numbers to determine the scale-up effects. Alternatively, a different unit operation may be chosen for which the design and scale-up information is available.

15.3 Example case Oscillatory Flow Reactor (OFR)

A fine chemicals company is considering to apply Process Intensified Unit Operations in general and is in particular interested in the OFR for a new multipurpose reactor, replacing their mechanically stirred tank reactor operated in batch. The company has four of these reactors, for scale-up purpose with volumes: 0.001, 0.04, 1, and 20 m^3.

They ask their recently recruited young chemical engineer to compare the two reactor types and make a recommendation.

Table 15.1: Oscillatory flow reactor performance characteristics.

Performance phenomenon	Characteristic
Mass transfer gas-liquid	Similar to mechanically stirred tank
Mass transfer liquid-liquid	Similar to mechanically stirred tank
Residence time	Long and similar to batch
Residence time distributions phases	Narrow and same as batch
Phase flow directions	Cocurrent and same as batch
Heat transfer performance to outer wall	Higher than batch
Impulse transfer to baffles	Higher than mechanically stirred tank
Three-phase application	Same as batch

She reads a review article [3] and then she makes Table 15.1. She recommends buying initially a $0.04\,\text{m}^3$ OFR each with 40 annular baffles.

She recommends running the OFR reactor for an existing commercial scale product and to compare the reactor performance with the existing commercial scale reactor. If the OFR performance is equally well or better, then a large scale OFR should be purchased, as the scale-up procedure will be greatly reduced, product quality control will be better, and the productivity of the unit is far larger than the batch reactor.

Bibliography

[1]   Green DW. Perry's Chemical Engineers' Handbook. 9th ed. New York: McGraw Hill; 2016.
[2]   Stankiewicz A, Van Gerven T, Stefanidis G. The Fundamentals of Process Intensification. Wiley; 2019.
[3]   Bianchi P, et al. Oscillatory flow reactors for synthetic chemistry applications. Journal of Flow Chemistry. 2020;10(3):475–90.

16 Distillation design

16.1 Basics distillation

Distillation is still the most applied separation technology in the oil and gas and chemical industry [1]. A rapid concept design method requiring no computer package is provided in this chapter. If that design is feasible and seems attractive, a flow sheet package can be applied to obtain an optimized concept design. In the end, it may be followed up by a detailed design using, for instance, Perry's Chemical Engineers' Handbook [1].

Distillation is based on differences between the vapor composition in equilibrium with the liquid composition at a certain temperature. For designing a distillation column, these equilibria have to be known over the whole temperature range. Preferably, these equilibria data are also known for a lower pressure and for a higher pressure than atmospheric pressure, so that an optimum design can be made.

Nowadays physical property data models predicting these vapor-liquid equilibria are available with process flow sheet programs. The Dortmund Data Bank provides numerous vapor-liquid equilibria data determined experimentally [2].

However, if polar components and hydrogen bonding components such as alcohols are involved, then these data model predications need to be validated experimentally.

Also, if the liquid has a high viscosity and or has non-Newtonian fluid behavior, then the distillation column design needs to be validated experimentally in a test column.

16.2 Distillation design method

16.2.1 Thermodynamic equilibrium to select distillation type

The starting point of any distillation concept design is the vapor-liquid equilibrium diagram. As distillation is based on differences between the vapor composition and the liquid composition at the same temperature.

Figure 16.1 shows on the left-hand side a two-component system vapor-liquid equilibrium with temperature at the vertical axis, obtained from de Haan [3]. The figure on the right-hand side shows the vapor-liquid equilibrium at constant temperature (and pressure). It is clear from these figures that component y is more volatile than component x. The equilibrium is a smooth rising line meaning that the top composition of the distillation can be very pure in component y. And it also means that a bottom composition with a high purity of x can be obtained.

Figure 16.2 shows various vapor-liquid equilibria curves where at a certain point the vapor and liquid curves have the same composition. This point is called the azeotrope. This means that this is the maximum product purity that can be reached in a single column. Distillation with such an equilibrium curve is called azeotropic distillation.

Design for normal distillation is treated in Section 16.2.2.

https://doi.org/10.1515/9783111203256-021

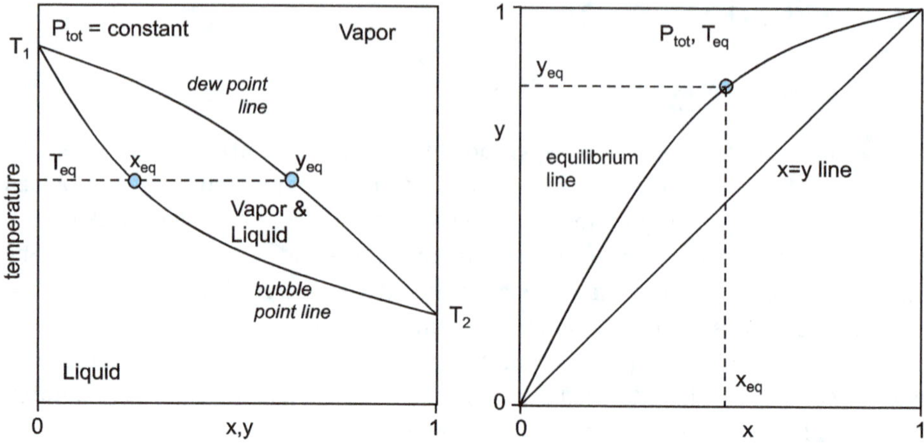

Figure 16.1: Two-components vapor-liquid equilibrium graphs.

Design for azeotropic distillations are treated in Section 16.2.3.

16.2.2 Mass transfer for normal distillation design

Binary systems with a normal distillation equilibrium curve (no azeotrope) can often be modeled with the distribution coefficient K:

$$K = y_i/x_i \tag{16.1}$$

where y_i is the mole fraction of component i in the vapor phase and x_i the mole fraction of component i in the liquid phase.

If two components indicated with i and j have to be separated, then a quick estimate can be made of the minimum number of trays that will be needed and the minimum reflux ratio with the following expressions.

First, a relative volatility parameter α_{ij} is introduced:

$$\alpha_{ij} = K_i/K_j \tag{16.2}$$

A rapid preliminary design can be made for simple distillation columns with a top and bottom outlet flow.

The minimum number of theoretical stages N_{min} can be calculated with the Fenske equation (16.3) [3]:

$$N_{min} = \frac{\ln\{[\frac{x_D}{1-x_D}]/[\frac{x_B}{1-x_B}]\}}{\ln \alpha_{ij}} \tag{16.3}$$

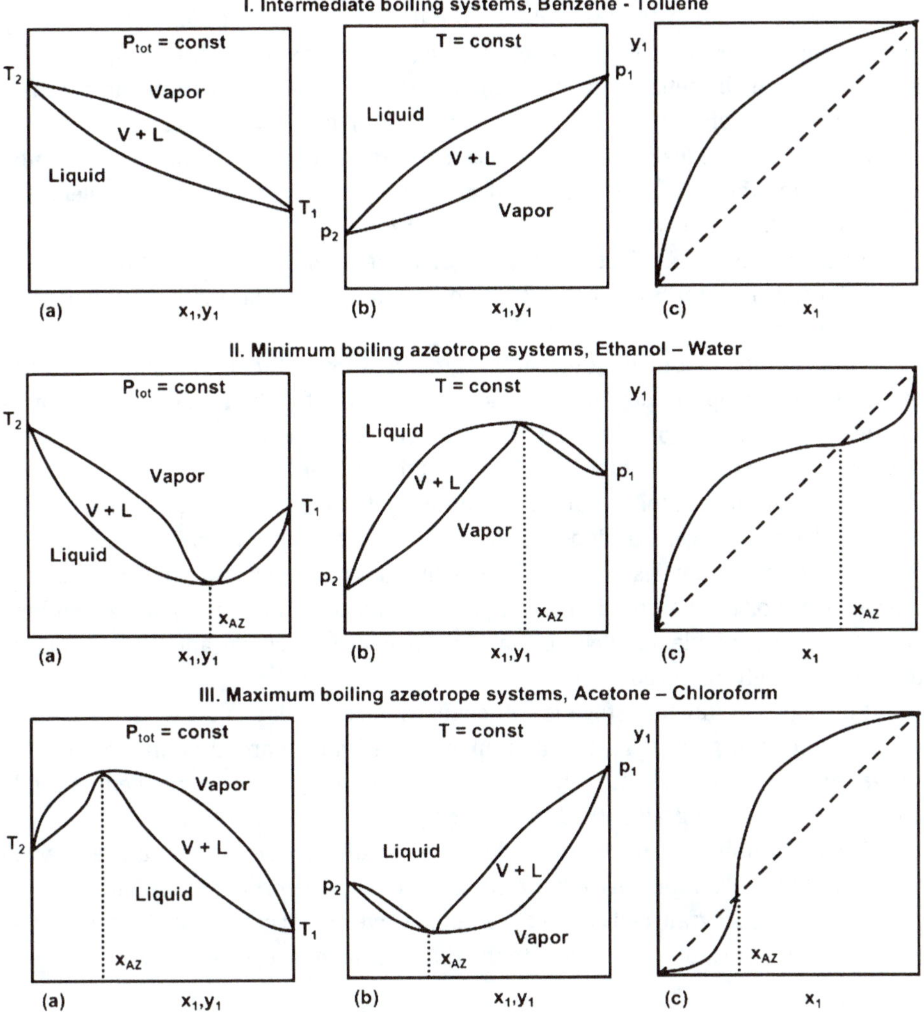

Figure 16.2: Two-component vapor-liquid equilibrium with azeotrope examples [3].

x_D: is the mole fraction of component i in the top of the column. So, it is the required purity of the top product stream for component i.

x_B: is the mole fraction of component j in the bottom product stream.

The minimum reflux ratio R_{min} is given by the Underwood expression (16.4) from [4]:

$$R_{min} = \frac{\frac{x_D}{x_F} - \alpha_{ij}\frac{1-x_D}{1-x_F}}{\alpha_{ij} - 1} \tag{16.4}$$

For the height of the distillation column, the design method of Chapter 17 can be applied, using N_{min} as input and then determining the Number of Transfer Units (NTU) by NTU = $2 N_{min}$ and by determining the Height of a Transfer Unit (HTU) with mass transfer coefficients. The height of the distillation column H is then H = NTU.HTU.

So, designing the distillation column can now start by taking the minimum number of plates using the Fenske equation and multiplying this by a factor of 2 to get a first reasonable estimate of the number plates required.

For the reflux ratio, a factor 1.5 higher than the minimum reflux ration according to the Underwood expression [xxx2.3] can be taken to get a reasonable estimate of the reflux required.

A multicomponent distillation can be simplified into a binary distillation problem by clustering the top components into a single component with properties of the lowest relativity of the cluster and by clustering the bottom components into a single component with properties of the highest relativity of the cluster.

The column height follows from the number of equilibrium stages and the height required for each stage. This depends strongly on the type of internals chosen, and the liquid transport properties and on whether the mass transfer limitation is mainly on the gas phase or on the liquid phase side. This subject is too large to be treated here. Textbooks are available on selecting plates or structured packings and then calculating the column height required.

If a single equilibrium stage is sufficient, then single flash vessel can be applied, where the feed mixture is first heated up, or cooled down, and then in the flash vessel, gas and liquid are separated by gravity. This phase separation can be enhanced by creating an enhanced gravity field by a cyclone.

If many equilibrium stages are needed, then the distillation column can be very tall, even over 100 m, Examples are ethane/ethene and propane/propene distillations.

A flow sheet program can be used to optimize investment cost, related to the number of trays and for variable cost related to the reflux ratio; hence, energy cost.

16.2.3 Residence time distribution considerations

Residence time distributions of the gas phase and the liquid phase flows are of course very important for the distillation column performance. Short-cut flows would cause poor distillation performance. Chapter 17 provides information on how a narrow RTD can be obtained.

16.2.4 Heat effects and heat transfer

Heat effects of distillation are heat of evaporation and heat of condensation. These local heat effects are directly compensated for by heat transfer coupled with the mass transfer of the components evaporating and absorbing. This can be understood because heat transfer between fluids is faster than mass transfer, due to the fact that the thermal diffusivity parameter ($\lambda/\rho C_p$) for fluids is orders of magnitude larger than molecular diffusion coefficients. See Chapter 22 for details of thermal and molecular diffusion comparison.

The energy required by the reboiler (Q) can be quickly estimated by expression (16.5) [3]:

$$Q = D(R + 1)\Delta H_{vap} \tag{16.5}$$

D: (mol/s) Top product flow
R: (-) Reflux ratio
ΔH_{vap} (J/mol) Evaporation enthalpy

16.2.5 Impulse transfer

Impulse transfer between the up flowing gas by pressure difference between bottom and top and the downflowing liquid by gravity force takes place through the column. When these two forces are equal, flooding occurs and liquid cannot flow downwards anymore. For all kinds of distillation, internal flooding correlations are available. The gas and liquid superficial velocities are important parameters in these flooding correlations. If for a chosen column diameter and packing type and size, flooding occurs, then this can be prevented by increasing the column diameter, by which the superficial velocities decrease. Also, the packing size can be increased to avoid flooding. An optimum column and packing design is in general made in the process development stage.

16.2.6 Azeotropic distillation

If the vapor equilibrium diagram shows azeotropic behavior and the desired product purity is beyond this azeotropic composition, then a normal design procedure cannot be followed. But a different design method has to be applied. Here are the four most applied methods for azeotropic distillation, treated with the same selection hierarchy as for molecular separation method selection of Chapter 15.

Investigate the first two connected classic distillation columns. One operating at a higher pressure (and temperatures) by which the azeotropic composition changes and the desired product purity of one product stream can be obtained in one of the columns; say, the one designed for a higher pressure. The remaining other stream is recycled back to the other lower pressure column where the desired second pure product stream is obtained [3].

This option should be considered first as it does not require an additional component, and each distillation column can be quickly designed, if it appears that the two-column solution requires many stages.

A process scheme for a two-distillation column butanol-water is presented in Figure 16.3 [5]. Distillation column 1 operates at a pressure different from distillation column 2. A common liquid-liquid phase separator is used to obtain a water-rich reflux and an alcohol rich reflux stream to each of the columns.

Now each distillation column can be designed using the binary design method.

If the pressure to obtain the desired separation is very high or very low, then a second separation option should be investigated.

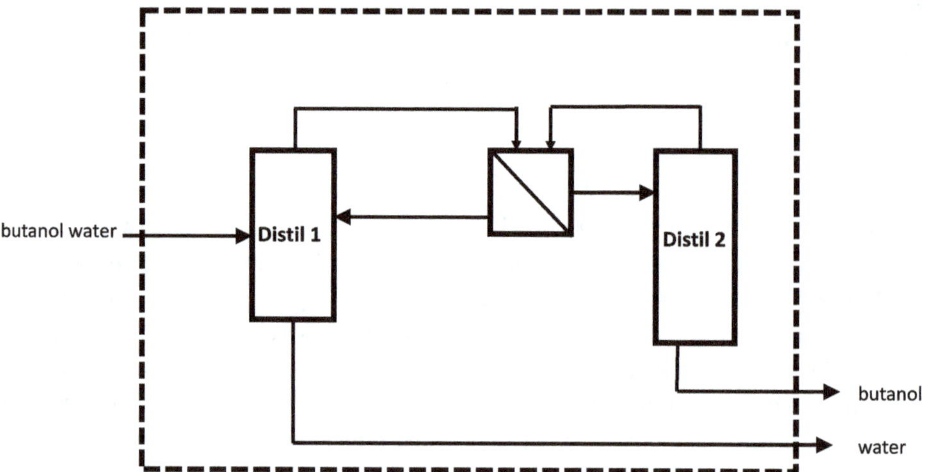

Figure 16.3: Butanol-water azeotropic two-column distillation.

Investigate second: A membrane in combination with a distillation column. The membrane then should produce the desired pure product stream for which membranes are available. The distillation produces the second component product stream. The distillation column also concentrates the first component in the remaining distillation stream. This second stream is then recycled to the membrane feed [6]. The process scheme is obtained by replacing the membrane of the first distillation of Figure 16.3.

This option should be considered as the second option as it still does not need a solvent, but only a membrane unit in combination with the distillation. The membrane design will however take more time. Chapter 20 can be helpful in making such a design.

If no membrane separation is available for this option, then a third option should be considered.

Investigate third: Add a component to the feed of the first distillation column. This component is often called an entrainer, or a solvent. This solvent changes the vapor-equilibrium curve such that the azeotrope disappears. The added component is recovered in a second distillation column and recycled back to the first column [3]. A good solvent selection is of course a very important part of this design, but that goes beyond the scope of this book.

Investigate four: Apply reactive distillation. The reaction removes one of the components so that a pure product stream can be obtained by simple distillation. This is of course only an option if a reaction is needed anyway in the process design. Lutze [6] provides concept design methods for this option.

16.2.7 Limitations of design method

The presented method design method for normal distillation only applies to ideal vapor-liquid equilibria with a linear relationship between vapor and liquid. Hence, nonideality behavior, and non-linear equilibria cannot be treated with this method, for those vapor-liquid equilibria elaborate models can be applied with flow sheet programs.

16.3 Example case: Depropanizer distillation design in oil refining

Gomez provides a nice distillation design study for training students about the separation of light alkane hydrocarbons methane, ethane, and propane (C1, C2, C3) from heavier hydrocarbons butane, pentane, and hexane (C4, C5, C6). It is called a depropanizer, as propane and lighter components are to be separated from butane and heavier components. Gomez also provides the distribution coefficients relative to C4 and also the design feed, top product, and bottom product compositions together with the feed, and product stream compositions [6]. The information is shown in Tables 16.1 and 16.2.

Table 16.1: Distribution coefficients and relative volatility alkane hydrocarbons.

	C1	C2	C3	C4	C5	C6
K_i	8.61	2.98	1.35	0.61	0.29	0.14
$\acute{a}_{iC4}$	14.0	4.86	2.20	1.00	0.47	0.22

By simplifying the separation problem into a binary separation problem of separating C3 from C4, the minimum number of theoretical stages can then be calculated from the Fenske expression (16.3). The minimum Reflux ratio R_{min} can then also be calculated.

This example shows that multicomponent distillation can in some cases be treated as a two-component system by clustering.

The design problem of separating 6 components is simplified into a binary separation problem. Components C1, C2, C3 are clustered into a single top mole component, with fraction x_D. The propane properties are taken to represent this cluster. The idea

Table 16.2: Components flow feed, top product, and bottom product.

Component	Feed kmol/h	Top product kmol/h	Bottom product kmol/h
C1	26	26	0.0
C2	9.0	9.0	0.0
C3	25	24.6	0.4
C4	17	0.3	16.7
C5	11	0.0	11.0
C6	12	0.0	12.0
Total	100	59.9	40.1

behind this is that propane has the lowest relative volatility to butane compared to C1 and C2, so is the hardest to separate from the heavy components C4–C6. The properties of components C4–C6 are clustered into a bottom mole fraction x_B. For this cluster, the properties of butane are taken for the same reason. So, now the problem is a binary distillation problem.

The minimum Reflux R_{min} and the minimum number of theoretical plates N_{min} can now be determined by expressions (16.4) and (16.5). The results are shown in Table 16.3.

Gomez shows various designs using a flow sheet package. An optimized design had 24 stages and a reflux ratio 1.3 times the minimum reflux ratio [7]. So, the optimized design number of stages is a factor 2 higher than minimum number of stages required. The optimized reflux ratio is a factor 1.3 higher than the minimum reflux ratio.

Table 16.3: Binary design case depropanizer expressed in design parameters of this chapter.

Parameter	Value
x_D	0.995
x_B	0.990
$\acute{a}_{ij}$	2.2
x_F	0.6
N_{min}	12
R_{min}	1.3
N	24
R	1.7

Bibliography

[1]   Green DW. Perry's Chemical Engineers' Handbook. 9th ed. New York: McGraw Hill; 2016.

[2]   Dortmund Data Bank. Sourced 27 June 2025. https://www.ddbst.com/ddb.html.

[3]   De Haan AB, Eral HB, Schuur B. Industrial Separation Processes: Thermal Unit Operations and Mechanical Unit Operations. Berlin: De Gruyter; 2025.

[4] Seider WD, Lewin DR, Seader JD, Widagdo S, Gani R, Ng KM. Product and Process Design Principles: Synthesis, Analysis and Evaluation. John Wiley & Sons; 2016.

[5] Di Pretoro A. Dynamic simulation and plantwide control of an ABE/W separation plant. MA thesis. Milan, MI: Politecnico di Milano. 2017 (see pp. 12, 27, 38).

[6] Lutze P, Gorak A, editors. Reactive and Membrane-Assisted Separations. Berlin: De Gruyter; 2016.

[7] Gómez-Siurana A, Font A. Exploration of the design of distillation columns—Beyond reflux and tray count. 2021. Sourced 2-10-2024. https://journals.flvc.org/cee/article/view/123523.

17 Absorption design

17.1 Basics absorption

The term absorption for physical separations is used in two ways. It is used for the top part of distillation columns where certain components from the gas phase are absorbed in the downflowing reflux liquid phase. It is also used for absorption columns to selectively absorb a component from the gas phase into a liquid phase. The purpose of this latter unit operation is to remove a component from the gas feed stream.

The column can also be to remove a component from the liquid stream to the gas stream, such as in deaeration of water. The latter is called desorption, or stripping. The design method for the latter is similar to absorption.

An absorption column has in most cases a dumped packing of particles, or a structured packing, to create a high gas-liquid interface area and high gas-liquid mass transfer coefficients. This packing also facilitates countercurrent gas and liquid flow by the high space outside the structured packing material. Finally, the structured packing narrows residence time distributions of the gas and liquid flows, by which deep removal of components is feasible in a limited column size.

Often a chemical substance is added to the solvent, which reacts with the component to be removed from the gas stream. For that case, the absorption column is a reactor and not a physical separation. A design method for such a gas-liquid reactor is described in Chapter 9.

17.2 Design method

17.2.1 Thermodynamic equilibria and solvent selection

Absorption of a component from the gas phase into the liquid phase is based on an equilibrium of that component between the gas and liquid phase. The equilibrium parameter is sensitive to the temperature. Physical property databases are the first source for finding information about the parameter. A check should be made whether the information has been experimentally validated.

The solvent to be selected for the absorption is very important because the resulting loaded solvent will have to be further processed. The solvent should have affinity with the component to be removed from the gas stream. The component may also be removable from the loaded solvent by distillation, or stripping.

Often, however, water is chosen as absorbent for removing trace components from a gas stream. The water loaded with the trace component is then further treated in the existing Wastewater Treatment (WWT) at the process site, but still it has to be checked whether this loaded water stream is compatible for that WWT. The solvent selection should be evaluated anyway with the downstream processing in mind.

https://doi.org/10.1515/9783111203256-022

Because of the many options available and a no elaborate selection method for industrial practice is available from literature, a detailed solvent selection method cannot be presented here.

17.2.2 Mass transfer of components from one phase to another phase

Absorption is executed in general columns or towers with dumped or structured packing. Many technology providers of structured packings have their own design methods. Some also have scaled-down packings for laboratory test units for which the scale-down and scale-up information on mass transfer and residence time distribution is available [1, 2]. We therefore focus on first concept design to make a first estimate of the column size.

If the equipment size should be small, for instance, on oil or gas platforms at sea, or submarine vessels, where space is expensive, or when the pressure to be applied is high, then a rotating structured packed bed absorber can be applied [3]. These have much higher mass transfer coefficients than packed columns, due to the much higher gravitational acceleration. This results in much smaller equipment [3].

Mass transfer from the gas phase to the liquid phase is described in Chapter 9. Here is a summary. The overall flux of A from the bulk of the gas phase to the bulk of the liquid phase J_A is given by Westerterp [4]:

$$aJ_A = a\{mk_l/(1 + (mk_l/k_g))\}(C_{Agbulk} - (1/m)C_{Albulk}) \tag{17.1}$$

The mass transfer coefficient's part is clustered in the Mass Transfer Performance (MTP) term:

$$MTP = a\{mk_l/(1 + (mk_l/k_g))\} \tag{17.2}$$

The concentration driving force for mass transfer is defined by ΔC,

$$\Delta C = (mC_{Agbulk} - C_{Albulk}) \tag{17.3}$$

With this, the mass transfer expression (10.1) is now rewritten as

$$aJ_A = MTP \cdot \Delta C \tag{17.4}$$

Proper concept design of absorption columns thus requires information on the mass transfer performance MTP. Hence, information on a, k_l, k_g and m is needed for the specific design case at hand. If those are available, then a design can be made.

The equilibrium value m in MTP, called distribution coefficient or partition coefficient, is defined as

$$m = C_{Aleq}/C_{Ageq} \tag{17.5}$$

This is a very important parameter in the mass transfer performance term MTP. For values of m around m = 1, the liquid mass transfer will then be the dominant term in MTP, as the liquid mass transfer coefficient k_l will be much lower than the gas mass transfer coefficient k_g, due to the fact that for the majority of cases, the liquid diffusion coefficient is a factor 1,000 lower than the gas transfer coefficient.

This parameter m is based on expressing the concentration of A in the liquid phase: C_{Aleq} and the concentration of A in the gas phase at equilibrium in mol/m^3. This is different from most textbooks of physical chemistry where physical equilibria are based on mol fraction expressions in each the two phases are on partial pressures such as in Henry's equilibrium law and Henry's coefficients. So, values of m often need a transformation calculation from Henri's coefficients to distribution coefficients. By using the gas law, this can be simply executed. The simplest form of the gas law PV = nRT can be used here for this conversion. So, m values can be obtained from the Henri coefficient H for the standard condition by Henri expressed in N/m^2 by

$$m = H/RT, \quad \text{with } R = 8.314 \text{ J/(mol K) and } T = 298 \text{ K}$$

The reader should be aware that the Henri coefficient is not always expressed in s. i. units, so an additional conversion factor may be needed.

Hegely provides a comprehensive overview of mass transfer coefficients, a, k_l, and k_g for various packing types [1]. The reader is referred to that article as a good entry to various correlations to calculate values for these 3 parameters for packing types, gas, and liquid velocities and properties. It is recommended to search for the most relevant correlations for the case at hand and then take the most conservative values obtained. With those data and with the equilibrium parameter m from physical basic data, values for MTP can be determined.

Table 17.1 contains some Henri coefficients for industrially relevant absorber design cases with water as absorbent at Temperature = 298 K. obtained from de Haan [5].

Table 17.1: Henri coefficients for absorption in water.

Gas component	Henri coefficients H 10^5 N/m^2 (bar)/mole
NH_3	2.0
SO_2	1.6
Cl_2	18
CO_2	67

Table 17.1 shows that the solubility of components in water varies enormously. This means that the value of m has to be obtained for a given design case, and thereby the

mass transfer performance parameter MTP. Copying a design column for a given gas from a different gas column design is not feasible.

De Haan also shows in a graph that Henri coefficients (hence, the distribution coefficient) decrease strongly with increasing temperature [5]. The distribution coefficient furthermore depends on other components in the liquid. Hence, mass transfer from the gas bulk to the liquid bulk is also affected by the temperature and actual components in the liquid.

The required packing height for a given degree of removal of the gas component from the gas flow can be calculated as follows. First, the minimum number of equilibrium stages; N_{min} is to be determined. For cases where the volumetric flow rates in the absorption column are nearly uniform over the column heigh, which holds for dilute gas streams, the Kremser equation can be used to determine N_{nun}.

The Kremser equation for N_{min} is mostly expressed in mole fractions. However, the expression can also be used if it is based on volumetric flows and molar concentrations. This can be simply done by replacing the mol fractions y and x by gas and liquid concentrations expressed in mol/m^3 and by replacing the equilibrium constant K by 1/m.

The Kremser equation is provided by De Haan [5]. The gas mol fractions and the liquid mol fractions and the mol flows are replaced by concentrations, the mol flow rates by volumetric flow rates, K by 1/m.

The Kremser equation is then given by

$$N_{min} = \frac{\ln\left[\frac{Cg_{in}-(1/m)Cl_{in}^{*}}{Cg_{out}-(1/m)Cl_{in}^{*}}\left(1-\frac{1}{A}\right)+\frac{1}{A}\right]}{\ln(A)} \quad \text{(for A ≠ 1)} \tag{17.6}$$

in which the absorption factor A is then defined as

$$A = \frac{mL}{G} \tag{17.7}$$

N_{min}: The minimum number of equilibrium stages
L: liquid flow (m^3/s)
G: gas flow (m^3/s)

For the component to be removed from the gas by absorption, the following concentrations are defined:
Cg_{in}: The gas concentration of the inlet (mol/m^3)
Cg_{out}: The gas concentration of the outlet (mol/m^3)
Cl_{in}: The liquid concentration of the inlet (mol/m^3)

The absorption factor A is the first key design parameter for designing the absorption column.

If A = 1, then there is no driving force for mass transfer and the column height according to the Kremser equation would turn out to be infinitely long. So, A should be chosen to be larger than 1.

If A = infinite, then the driving force of mass transfer is at its maximum, which means that the concentration of the component is near zero. For plug flow of the gas phase, the Rmass transfer relation then reduces to

$$C_{gasout}/C_{gasin} = e^{-NTU}$$

With $NTU = MTP . H/v_{sg}$.

So, in absorber design, first the value of A has to be chosen. A should be larger than 1; hence, a surplus of absorbent relative to the minimum absorbent at A = 1 is needed. The value of A not only affects the size of the absorber, but also the cost of downstream treatment of the loaded absorber liquid.

A starting point for the absorber sizing can be to choose A = 2. The minimum number of transfer units N_{min} follows then from equation (17.6).

The Number of Transfer Units (NTU) should be larger than N_{min}. Skolund showed that the following relationship provides a good comparison between designs based on the number of equilibrium stages and designs based on mass transfer models with gas and liquid flow in plug flow [6]:

$$NTU = 2 N_{min} \tag{17.8}$$

The Height of a Transfer Unit (HTU) can be approximated by

$$HTU = v_{gsup}/MTP \tag{17.9}$$

The total packing height H can now be calculated by

$$H = HTU . NTU \tag{17.10}$$

The column diameter can now be determined from the total gas flow and the superficial gas velocity v_{sg}, which is suitable for the packing to obtain a reasonable mass transfer coefficient and no flooding.

An ultrarapid short-cut absorber design

A rapid short-cut method for an absorber design can be made by taking a large surplus of absorber liquid so that the driving force concentration difference for mass transfer for the gas component has its maximum everywhere in the column.

The same short-cut column approach can be obtained by adding a reactive chemical to the absorbing liquid, such as an alkaline additive in case of acid gas absorption. This additive will reduce the bulk liquid concentration to zero, so that maximum concentration differences are obtained over the whole column length.

The number of plug flow mass transfer stages $NTU = MTP \cdot H/v_{sg}$ then given by the gas concentration ratio outlet to inlet:

$$C_{gasout}/C_{gasin} = e^{-NTU} \tag{17.11}$$

With

$$NTU = MTP \cdot H/v_{sg} \tag{17.12}$$

So, the height of the column follows from

$$H = v_{sg}NTU/MTP \tag{17.13}$$

For 95 % removal, $C_{gasout}/C_{gasin} = 0.05$; $NTU = 3$.

Choose as a starting point a typical value for $v_{sg} = 1\,m/s$. Find values for k_g, k_l, and m and a packing height value for H is determined. The superficial gas velocity and the actual gas feed flow rate determine then the required column diameter.

In this way, also various packing sizes can be quickly compared for their effect of column height H.

17.2.3 Residence time distribution effects

Residence time distribution of the gas phase and the liquid phase is also an important aspect of absorption column designs. For properly designed packed beds with a good liquid distribution at the top and for commercial scale sizes, the Reynolds numbers for gas and for liquid should be

$$Re > 10$$

So that

$$Bo = 2$$

The overall Peclet number; given by

$$Pe = Bo\,L/d_p \tag{17.14}$$

then shows that the desired Plug flow behavior can then be easily reached as follows.

Determine the minimum Pe value Pe_{min} by using the removal degree in the same way as for conversion X in Chapter 9. For most cases, plug flow will be easily obtained by designing the column length L. The description for mass transfer is then sufficient for designing the column.

Warning about use of a Height Equivalent of a Theoretical Plate (HETP)

An alternative method for calculating the height of the packing is using the concept of the Height Equivalent of a Theoretical Plate (HETP). HETP values can be obtained from tables on packing types and their HETP. The actual height of a packing can then be obtained by multiplying the (minimum) number of equilibrium stages with the HETP value.

Here is my story on why this HETP approach is to be avoided. In the years I have worked as advising technologist at Shell Chemicals site in Pernis, I have had two cases where an absorption column for the removal of a trace component from a gas stream did not reach the desired degree of separation. In both cases, an acid gas component was to be absorbed in water. The technologists of those chemical plants had checked the absorption column design method, based on the Height Equivalent of a Theoretical Plate (HETP) values and could not find an error. So, they came to me.

I checked the experiments on which the HETP was based, and it appeared that they were based on distillation experiments with hydrocarbons, which have high value for the equilibrium parameter m. In fact, the m values were so high that the mass transfer was mainly limited by the gas phase side, so the HETP mainly represented $k_g a$ values.

For the design cases at hand, however, the equilibrium values for m were much lower and the main resistance to mass transfer was on the liquid side. Liquid side mass transfer values of k_l are in general much lower that k_g values because the diffusion coefficients for liquids are a factor 10^4 lower than for gases, so k_l values are always much lower than k_g values.

The background of the HETP approach is that, if the equilibrium value of m is so high that the mass transfer is limited by the gas phase side, and not on the side of the gas-liquid interface, then the mass transfer coefficient k_g is governing the mass transfer. This k_g is increasing with increasing superficial gas velocity. For a limited gas velocity range, the ratio $k_g a / v_{gsup}$ is then nearly a constant value with the dimension meter. As long as the value of m, the gas density and the gas viscosity of the design application are near the gas properties for which the HETP is determined, a reasonable design estimate can be made with this HETP approach. HETP values for various packing are reported, so a rapid concept design can be made. In all other cases, the HETP approach will give the wrong design and often a too short column design.

17.2.4 Heat effects and heat transfer

Heat effects of absorption are the same as in distillation, As however the component in the gas phase to be absorbed into the liquid phase is present at low concentrations, the heat of absorption often can be neglected. A simple maximum temperature rise can be estimated using the released heat by absorption and the heat capacity of the gas and of the liquid phase. If the temperature rise is less than 1 °K, it can be neglected.

The maximum temperature rise by absorption heat can be estimated from

$$\Delta T_{abmax} = (\Delta H_{ab} \cdot M_{ad})/(F_l \cdot Cp_{fl} + F_g \cdot Cp_g) \tag{17.15}$$

ΔT_{abmax}: (K) Maximum temperature rise fluid by absorption heat
F_l: (kg/s) Liquid feed flow rate
F_g: (kg/s) Gas feed flow rate
Cp_l: (J/(kg K) Specific heat capacity liquid feed flow
Cp_g: (J/(kg K) Specific heat capacity gas feed flow
ΔH_{ad}: (J/(mol)) Heat of adsorption per mol of absorbing component
M_{ab}: (mol/s) Absorption component feed flow rate

There is a second heat effect to be considered and that is that the gas phase and the liquid phase feed temperatures feeds can differ. If there are temperature differences between the two feeds, then first an estimate has to be made what the average temperature in the column will be assuming an instant heat exchange rate.

The temperature difference over the length of the column can be estimated by a simple heat balance over the gas phase and the liquid phase. By assuming maximum heat transfer between gas and liquid inside the column the gas temperature at the gas outlet will be equal to the liquid feed input; hence, a maximum heat transfer rate is assumed. Furthermore, for the simplest heat balance no heat of absorption and no heat of evaporation is taken into account.

The liquid temperature difference between inlet and outlet can then be estimated from

$$T_{lout} - T_{lin} = (F_g C_{pg}/F_l C_{pl})[T_{gin} - T_{lin}] \tag{17.16}$$

T_{lout}: liquid outlet temperature (K)
T_{lin}: liquid inlet temperature (K)
T_{gin}: liquid outlet temperature (K)
F_g: gas mass flow (kg/s)
F_l: liquid mass flow (kg/s)
C_{pg}: specific heat coefficient gas (J/(kg.K))
C_{pl}: specific heat coefficient liquid (J/(kg.K))

A conservative value of the partition coefficient for the liquid can now be chosen by taking the lowest value for the inlet and outlet liquid temperature as representative for the whole column. For most cases, the temperature rise of the liquid will be in the order of a few degrees Kelvin. The case of Section 17.3 shows that even for a gas inlet temperature of 120 °C and a liquid temperature of 20 °C the liquid temperature rises only 1 °C.

17.2.5 Impulse transfer and flooding

Flooding means that the pressure drop by the up-flow gas phase is a larger up-flow force than the gravity force by the down-flowing liquid, so that operation is not possible. This means that gas and liquid velocities are limited by this flooding phenomenon, and thereby packing heights are limited by flooding. By increasing the column diameter, the velocities can be lowered, and so flooding can be avoided.

Flooding correlations for packings involving gas and liquid flow properties are available so that the maximum gas and liquid superficial velocities can be determined for each packing. In this way, a reasonable packing and column dimension design can be made.

17.2.6 Limitations of design method

The first limitation of the design method is its use of the Kremser equation. This equation is only valid for the case that:
- Both the operating and equilibrium lines are straight.
- The liquid and gas flow rates) are constant throughout the column, which is a reasonable assumption for dilute systems.
- A linear equilibrium law applies.

A second limitation is that that column design optimization for pressure a variable is not treated. As compressors are expensive, it means that for most cases the pressure of the absorber is chosen is the pressure of the feed gas.

17.3 Example case: Absorber design SO_2 removal from flue gas

Rachel is a young chemical engineer, who starts her career at a specialty chemicals company. The company produces specialty chemicals from agricultural feedstocks. The chemicals plant stands in a small chemicals production site, but has its own furnace, fired by gas oil, for generating steam for various duties. Due to stringer flue gas emission regulations, the acid gas content in the flue gas has to be reduced by a factor of 20 (95 % removal). Rachel is asked ast her first job to make a design to get this reduction.

Rachel is reading this book and generates some options for this reduction. One of the options is absorbing the acid components in water by a column absorber. The main component to be absorbed is SO_2. She searches for an absorber design and finds a paper by Harry-Ngei [7]. The paper describes an absorber for SO_2 using water as absorbent and it provides correlations for mass transfer.

Barbara reads this book and then decides first to find a destination for the resulting salty water stream. She does not want the salty water stream to end in the river, which is probably not allowed anyway. She asks around inside the company, and also outside the company, using the chapter on circular economy. She finds a nearby company, which has special unit for converting salty water into demi-water and solid potassium sulfite. The latter is sold to a company. The demi-water is used for steam generation. She asks whether her salty water stream containing sulfite can be fed to this unit. The answer is "yes" if the water stream is less than $0.2\,\mathrm{m}^3/\mathrm{s}$.

With this information, she first calculates the absorber factor A using this water flow rate as input. Table 17.2 contains the value of A and all other parameters for the absorber column design and also all parameters used by Harry-Ngei [7]. It appears that the value of A is 15 % lower than the value used by Harry. So, the column height will be longer than that of Harry's design.

However, she then considers using alkaline in addition to the water feed, so that the SO_3 concentration in the water drops to zero, because of its rapid reaction with alkaline. It appears that if she used KOH as alkaline, then the resulting aqueous KSO_3 stream is acceptable for the waste water unit.

She decides first to calculate the temperature rise of the water inside the column using the heat balance calculation. The flue gas has a temperature of 398 °K and the demi-water has a temperature of 298 K. The maximum water temperature rise calculated from 17.12 appears to be 1 °K. So, she takes the water temperature inside the column as 298 K, as for that temperature she has the Henri coefficient value. The Henri distribution coefficient for SO_2 in water and air at 298 °K is obtained from de Haan [5]. This results in the partition coefficient m = 64.5.

She then uses the Harry-Ngei packing parameter values, the correlations for mass transfer coefficients for a, k_l, and k_g and calculates the required packing height using expressions of Section 17.2.

She then checks whether the residence time distribution of gas is sufficiently narrow to fulfill the requirement of plug flow, the minimum Peclet number of Chapter 9, and a correlation for a trickle-bed Peclet of Chapter 10. The calculated result is shown in Table 17.2. Plug flow is indeed achieved.

She writes an executive summary for her management on the column sizing and the wastewater destination at the neighboring plant.

She recommends to go to an EPC contractor, absorber column technology provider. They should have mobile small absorber column to test the real flue gas absorption effectiveness and also to test corrosion rates of the construction material and its welding. That technology provider does the tests for free if they are chosen to be the EPC contractor for the column and the transport pipeline for the wastewater to the neighbor wastewater unit.

Her overall conclusion regarding the design methods on circular economy is that it widens the whole design problem to look for integral solutions. The absorption column design method is a practical concept design method, facilitating literature correlations on mass transfer to be used and provides a first estimate of column size and packing

Table 17.2: SO_2 flue gas water absorber design.

Parameter	Dimension	Design	Reference
Flue gas feed flow rate	kg/h	40,000	[7]
Fue gas feed flow rate	kg/s	11	
Flue gas feed rate (gas density 0.9 kg/m^3)	m^3/s	12	
SO_2 feed flow rate	kg/h	20	[7]
Temperature flue gas feed	°K	398	[7]
Specific heat coefficient gas	J/(kg°K)	1,000	
SO_2 partial pressure feed	mmHg	59	
SO_2 mol fraction feed gas	–	0.0078	[7]
SO_2 mol fraction outlet gas	–	3.9 10^{-4}	
SO_2 removal on feed	%	95	
Water feed flow rate	m^3/s	0.20	
Water feed flow rate	kg/s	200	
Temperature water feed	°K	298	
Temperature rise water	°K	1	
Specific heat coefficient water	J/(kg°K)	4,100	
Packing Pall ring diameter	M	0.055	[7]
Column diameter D	M	2.5	[7]
Cross sectional column area	m^2	5	
v_{ls}	m/s	0.040	
v_{gs}	m/s	1.7	
Specific surface area a	m^{-1}	70	[7]
liquid side mass transfer coefficient k_l	m/s	0.00017	[7]
Gas mass transfer coefficient k_g	m/s	0.02	[7]
Henri coefficient H	N/m^2	1.6 10^5	[5]
Distribution coefficient m	–	64.5	
MTP	1/s	0.5	
NTU	–	3	
Height of packing	m	10	
Removal degree X	–	0.95	
Pe_{min}	–	60	
Bo	–	2	
H/dp		2,000	
Pe	–	4,000	

size. The minimum height calculation for zero concentration in the liquid bulk by using an alkaline solution is a useful piece of information.

Bibliography

[1] Hegely L, Roesler J, Alix P, Rouzineau D, Meyer M. Absorption methods for the determination of mass transfer parameters of packing internals: A literature review. AIChE Journal. 2017;63(8):3246–75.

[2] Flagiello D, Parisi A, Lancia A, Di Natale F. A review on gas-liquid mass transfer coefficients in packed-bed columns. Chemical Engineering. 2021 Aug 2;5(3):43.

[3] Stankiewicz A. The principles and domains of process intensification. Chemical Engineering Progress. 2020 Mar 1;116(3):23–8.

[4] Westerterp KR, Van Swaaij WP, Beenackers AA. Chemical Reactor Design and Operation. 1984.

[5] De Haan AB, Eral HB, Schuur B. Industrial Separation Processes: Thermal Unit Operations and Mechanical Unit Operations. Walter de Gruyter GmbH & Co KG; 2025.

[6] Skowlund C, Hlavinka M, Lopez M, Fitz C. Comparison of ideal stage and mass transfer models for separation processes with and without chemical reactions. In: Proceedings of Gas Processors Association. 2012.

[7] Harry-Ngei N, Ujile AA, Ede PN. Absorber sizing and costing required to control SO_2 emission from a combustion system. European Journal of Engineering and Technology Research. 2019 Dec 1;4(12):1–5.

18 Liquid-liquid extraction design

18.1 Basics liquid-liquid extraction

Liquid-Liquid Extraction (LLE) is a molecular separation method based on selective extraction of a component from one liquid phase into a second liquid phase. This second liquid phase is often called a solvent. After the extraction, a second separation step is needed to recover the solvent from the extractant. This is often a distillation step. The solvent is then recycled back to the extraction step.

LLE separation method is chosen if some other separations such as distillation or absorption are not feasible; see Chapter 15 on molecular separation method selection.

In some cases, reactive extraction is applied by which the extracted component reacts in the solvent, so that a maximum driving force is obtained for the component to be removed, facilitating deep removal in a limited sized extraction step. For that, extraction is to be treated as a reaction design case [1].

18.2 Liquid-liquid extraction design method

18.2.1 Thermodynamic equilibria, component selection, and solvent selection

18.2.1.1 Thermodynamic equilibria

Extraction of component from one liquid phase into the second liquid phase is based on an equilibrium of that component between the two phases. That equilibrium parameter is also sensitive to the temperature. Physical property databases are the first source for finding information about the parameter. A check should be made whether the information has been experimentally validated.

No generic solvent selection method can be provided here. The solvent should of course not be toxic. A few safe solvents with affinity for the component to be extracted should be chosen and a preliminary integral process design should be made using distillation to recover the solvent and the extracted component. The solvent with the lowest energy cost can be selected

18.2.1.2 Component selection to be extracted

The first question to answer is which component in the feed mixture should be extracted by the solvent. Is it the desired product component, or the undesired impurity component? Strangely enough, this question is not explicitly treated in textbooks, but only implicitly. De Haan, for instance, provides a list of industrial LLE applications [2]. From this list, it is clear that the desired pure product is selectively extracted by the solvent from a feed containing other components. So, the desired pure product is selectively solvent extracted from the feed mixture.

https://doi.org/10.1515/9783111203256-023

The other option—extracting the other components from the feed— is not even considered in textbooks. A plausible reason for this is that the chance of finding one solvent for extracting selectively several components and not the product is very low, and engineers do not want to waste time on options with a low chance of success. This then leads to the following guideline.

Guideline 1. If a pure product is to be obtained from a feed containing several other components, then select an extraction solvent for that product component.

However, if there are only two components to be separated from each other, then it is better to select the component of the lowest fraction in the feed mixture. Because that in general will mean that the downstream process steps of the LLE will be small. This then leads to the next guideline.

Guideline 2. If a small amount of one undesired component is present in the feed, then it is better to select an extraction solvent for that undesired component.

Guideline 2 means that the amount of solvent and the solvent recovery section will be much smaller than when the product is extracted from the undesired component.

Guideline 3. Check the consequences of the solubility of the extracting solvent in the feed stream. Often this solubility is so high that also solvent has to be recovered from the extracted stream leaving the extraction unit.

18.2.1.3 Solvent selection

The second question to answer is which solvent to select. There are various sequences to select a solvent. One can first generate a list of solvent options from industrially available solvents with affinity for the component to be extracted. Then perform lab-scale equilibrium experiments. Or calculate equilibrium values using physical properties models of a flow sheet program and then perform lab-scale equilibrium experiments.

This will result in a short-list of solvents. Then this short-list can be further reduced by applying selection criteria. The selection criteria can be split into hard criteria which must be adhered to and soft criteria, which can be used to optimize the solvent selection by trade-off reasoning and economical evaluations.

De Haan provides a long list of criteria for solvent selection [2]. He does not provide a hierarchy of these criteria. However, using a hierarchy of general process concept design criteria, the following hierarchy is established for hard and soft criteria.

Hard criteria for solvent selection

Hard criterion 1. Safety (Flammability), health (toxicity), and environmental (SHE) properties of the solvent have to be known.

Hard criterion 2. SHE properties have to be acceptable for use in commercial process operation.

Soft criteria for solvent selection

The soft criteria are derived from de Haan [2] and put a hierarchy of high importance to low importance. The hierarchy is based on economic heuristics treated in Chapter 5 and is based on correlation by Bridgewater [3] showing that the size of flow rate (ton/h) through the process step determines the investment cost of a process step.

Soft criterion 3 (A high selectivity for the product component). The selectivity of the solvent for the product component relative to other components determines the final product purity of the LLE process. If that purity cannot be obtained by LLE, then an additional process step, for instance, adsorption will be needed to obtain the desired product purity, adding to the investment and energy cost. So, this criterion is the most important soft criterion.

Soft criterion 4 (A high distribution coefficient for the component to be extracted). This criterion results in a low amount of solvent needed for the extraction. It thereby also results in a small downstream solvent recovery process step with low investment and energy cost.

Soft criterion 5 (Easy separation of solvent from product). The preferred separation method for the solvent recovery from the product is by distillation. So, a large boiling temperature difference between the product and the solvent is required.

If trace amounts of heavy boiling component are coextracted, then a heavy boiling solvent is preferred. Because then the product purity in this distillation step can be enhanced. The heavy boiling components can be removed by the solvent recycle back to the extractant. The heavy components in the solvent may prevent extraction from the feed and the heavy components can also be removed from the LLE process by a solvent bleed stream.

Soft criterion 6 (Low solubility of solvent in the feed (effluent) stream). The solvent will dissolve to some extent in the feed (effluent) stream. A low solubility reduces the cost of the solvent recovery step from this effluent stream, shown in Figure 18.1.

Soft criterion 7 (A high-density difference between solvent and feed). This criterion allows high superficial velocities of feed and solvent; hence, allows for a low column diameter, which reduces the investment cost.

Soft criterion 8 (Low solvent viscosity). For a low solvent viscosity, high mass transfer coefficient values are obtainable. It also facilitates rapid settling of the solvent loaded with product component, by which this settling sometimes can be accomplished in the extraction column, so that a separate settling tank is not needed.

Soft criterion 9 (Optimal interfacial tension). A very low interfacial tension between solvent and feed means that small droplets can be formed. A high interfacial area is then easily created in the extraction. This in turn means high mass transfer coefficients, so a lower column size.

However, low interfacial tension also means that droplet coalescence may be slower or even absent if an emulsion is formed.

18.2.2 Process design for mass transfer

18.2.2.1 integral process concept design

First, an integrated process concept design is to be made as shown in Figure 18.1 In the extractor, the feed is contacted with the solvent. For deep recovery of the product from the feed stream, a countercurrent extraction is the preferred functional concept.

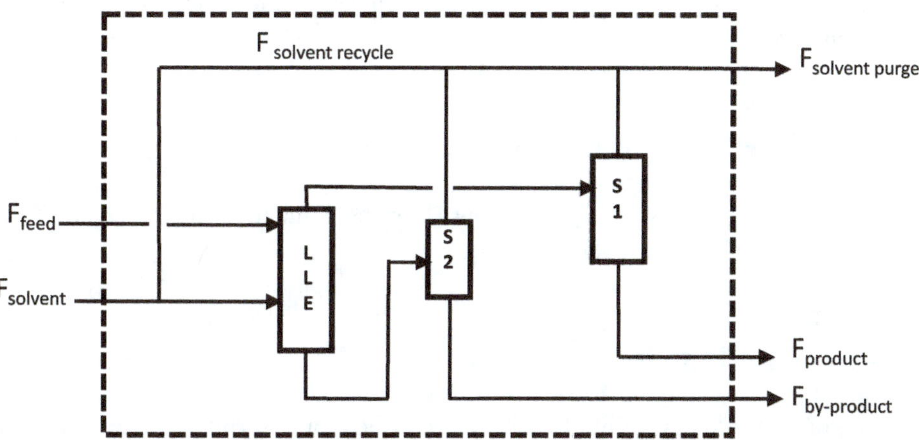

Figure 18.1: Liquid-liquid-extraction process flow scheme.

18.2.2.2 Column design for mass transfer

The preferred equipment for this functional step is a countercurrent packed bed column. The stream with the lowest density, most often the solvent feed will be fed to the bottom of the column. The stream with the highest density will be fed to the top of the column.

The sizing of this column can be made similar to distillation column design. First, the number of theoretical stages is determined. Then mass transfer coefficient correlations may be found in literature aan] (please review highlighted area) and the column can be sized. A good textbook on solvent extraction is by Lo [1].

The solvent loaded with the product from the extractor is sent to a recovery function block S1, which separates the product from the solvent. The solvent is recycled back to the extractor.

The second outlet stream from the extractor will contain some of the solvent. That solvent will be recovered in separate a solvent recovery step S2.

For low viscosity liquids with rapid coalescence rates, a structured packed bed in countercurrent flow is a low-cost option.

If, however, coalescence rates are low a separate coalescence tank can be added to the flow scheme.

The solvent recovery from the feed effluent and the solvent recovery from the loaded solvent should preferably be obtained by distillation.

18.2.3 Residence time distribution of fluids

The liquid flow residence time distributions should be sufficiently narrow to achieve the extraction degree in a limited column size. The same design method as for absorption can be chosen.

18.2.4 Heat effects and heat transfer

Heat effects in extractions are in general low. The same estimate methods for heat effects in absorption can be applied for extraction.

18.2.5 Impulse transfer and flooding

Because of countercurrent flow, flooding as described for absorption can also occur in LLE. Reliable flooding correlations for packings, however, are not easily available. If a certain packing is chosen for the desired mass transfer and also a flooding correlation is found, then it is advised to stay far away from the flooding limit, say a factor 2 in velocities and confirm empirically the design in the development stage. The experiments can often be carried out at the pilot plant facilities of the technology provider for the column and packing.

18.2.6 Limitations of design method

LLE extraction design has many uncertainties. Because of these uncertainties concept design has a limited reliability, and experimental validation will be needed.

Here is a short list of uncertainties:

Uncertainty list LLE
U1: Uncertainty of accuracy liquid-liquid equilibrium models
U2: Uncertainty of dispersion behavior

U3: Uncertainty of coalescence behavior
U4: Uncertain mass transfer coefficients
U5: Unexpected chemical reactions in any of the process steps
U6: Concept design uncertainty

Ad U1: Equilibrium uncertainty

Equilibrium models of liquids are in general unreliable due to the many types of physical molecular interactions, such as dispersive, induced dipole, permanent dipole, H-bridges, complex formation, and ionic-ionic.

The equilibrium models can be used for a preselection of solvents. Equilibrium experiments will be needed to confirm and adjust the equilibrium values. An experimental set-up is provided by Gladz [4].

Ad U2: Uncertainty dispersion behavior

The droplet dispersion is for nearly all equipment initiated by a distributor. The size of the droplet depends on the velocity of liquid through the holes, the interfacial tension, and the liquid viscosities.

However, coalescence of droplets forming bigger droplets can occur inside the column. This coalescence can also be influenced by interface active components with a preference for the interface. These components may be present in small amounts in the feeds.

This uncertainty will also affect the flooding behavior of the liquid flows.

Ad U3: Uncertainty of coalescence behavior

After the extraction, the dispersed phase has to be separated from the continuous phase. This can be achieved by creating an environment of no turbulence in which the dispersed droplets coalesce to big droplets and move by gravity difference into a second homogeneous continuous phase. Coalescence benefits from a high interface tension and low liquid viscosities of both phases. However, trace amounts of interface loving components can hamper the coalescence, so that an emulsion is created.

This uncertainty will also affect the flooding behavior of the liquid flows.

Ad U4: Uncertain mass transfer coefficients

The overall mass transfer coefficient depends on film coefficients on both sides of the interface: k_{l1}, and k_{l2} and on the interfacial area a. All three coefficients are first of all strongly affected by the droplet size; hence, by the dispersion behavior, and second by the viscosity, diffusion coefficients, and on third, by the liquid velocities of both phases.

Ad U5: Unexpected chemical reactions in any of the process steps

Unexpected and undesired reactions can occur in the LLE process due to the additional residence time in the three process steps, by the contact of the solvent with the feed, and by the higher temperature in the distillation columns.

Ad U6: Process concept design uncertainty removal

Process design uncertainties U1–U6 can be removed by an experimental mini-plant. Mass transfer correlations for liquid-liquid extraction in a structured packed column can be validated by having a mini-plant and testing the extraction with the real feed and the industrial solvent for a range of feed and solvent flow rates. The product recovery and the solvent recovery process steps should also be included in the mini plant. So, that the fate of trace components and of undesired chemical reactions can be observed and counteractive measures can be taken.

Because of these uncertainties experimental validation of the liquid-liquid extraction concept design is essential.

18.3 Example case: Industrial case acetic acid extraction from water

Gladz describes the separation of acetic acid from water by Liquid-Liquid Extraction (LLE). The feed contains 30 % (m) of acetic acid. The extraction solvent chosen is ethyl acetate [4].

Ethyl acetate is a well-known bulk chemical. Garner provides physical data for the extraction [5].

Api concludes that there are no safety issues with the use of ethyl acetate. So, the hard criteria for solvent selection are fulfilled [6].

Gladz does not describe the selection procedure leading to the solvent ethyl acetate. But from their process design, it is clear that a feasible process design is obtained. The selected solvent ethyl acetate can be separated by distillation from both the product and from the water effluent stream.

The function process concept is the same as shown in Figure 18.1.

The actual extraction step is carried out in a countercurrent packed bed.

The product separation from the solvent function block S1 is carried out in a distillation column, where the product leaves the top of the column and the solvent at the bottom.

The solvent recovery from the water effluent stream function block S2 is carried out as a distillation column, called a stripper. After the reflux condenser the solvent is sent to a hold-up tank. This solvent hold-up tank is common for both distillations.

Bibliography

[1] Lo TC, Baird MH, Hanson C. Handbook of Solvent Extraction. New York: John Wiley & Sons; 1983 (reprinted 1991).

[2] De Haan AB, Eral HB, Schuur B. Industrial Separation Processes: Thermal Unit Operations and Mechanical Unit Operations. de Gruyter; 2025.

[3] Bridgewater AV. The functional unit approach to rapid cost estimation. The Cost Engineer. 1974;13(5).

[4] Gladz D, Cross B. Designing liquid-liquid extraction columns. An Introduction, AIChE—East Tennessee Local Section, December 8th, 2020. Sourced: 16 July 2020. https://kochmodular.com/liquid-liquid-extraction/.

[5] Garner FH, Ellis SR, Pearce CJ. Extraction of acetic acid from water: 3—Binary vapour-liquid equilibrium data. Chemical Engineering Science. 1954 Apr 1;3(2):48–54.

[6] Api A, Belsito D, Botelho D, Bruze M, Burton G Jr, Cancellieri M, Chon H, Dagli M, Date M, Dekant W, Deodhar C. RIFM fragrance ingredient safety assessment, ethyl acetate, CAS Registry Number 141-78-6. Food and chemical toxicology : an international journal published for the British Industrial Biological Research Association. 2022 Sep 1;167:113363.

19 Adsorption design

19.1 Basics adsorption

Adsorption is applied for the removal of compounds from a liquid or gas stream, where other molecular separation fails. It is, for instance applied to remove trace amounts of organic compounds from groundwater in the preparation of drinking water, or for the separation of carbon dioxide from steam methane reforming reactor product stream to produce pure hydrogen.

One of the compounds to be separated from the others must have some affinity to the adsorption material, while the other components must not have this affinity, or a far lower affinity. This affinity separation is then achieved by feeding the mixture over a packed bed of porous adsorption material. The adsorbed component is removed from the stream, and the purified product leaves the packed bed.

In general, there is an equilibrium between the concentration of components in the feed and the concentration on the adsorbed material. In a packed bed, which for some time has been fed, three zones can be distinguished. A packed bed zone loaded with the adsorbed component in equilibrium with the component in the fluid at the feed concentration, a second zone where material is being adsorbed, so that the component in the fluid drops along the bed length, and a third zone where the packed bed does not contain adsorbed material and the fluid is free of the component. The middle zone moves slowly through the packed bed, until it reaches the end of the bed. Until that time, the product flow is nearly pure. When the concentration of the component to be removed increases at the outlet concentration, the feed to the bed is stopped and then fed to a second bed so that the separation is continued.

The length of the second zone where the concentration of the impurity drops determines the adsorption efficiency. If this zone is small, the bed can be nearly filled completely with adsorbed component. So, the design method focuses on the length of this zone and on the length of the entire bed.

The loaded bed is regenerated by desorption. This desorption can be obtained by heating the packed bed, called thermal swing regeneration, or by lowering the gas pressure, called pressure swing regeneration. For most cases, the adsorption stem is the slowest of the two steps, and thereby governs the bed size needed. For this first concept design, only the adsorption step is therefore designed, as this is the critical step for removing the component from the feed stream and it is also in general the slowest of the two steps: adsorption and desorption. So, we focus for the bed design on the adsorption step.

https://doi.org/10.1515/9783111203256-024

19.2 Adsorption bed design method

19.2.1 Thermodynamic equilibria

The separation by adsorption is based on selectively adsorbing a component from the feed stream by the porous adsorption material. The adsorption affinity for the component can be found by literature searches. Laboratory scale experiments can then be used to determine the so-called adsorption isotherm. This isotherm shows the equilibrium concentration of the component to be removed between the fluid and the adsorbent.

19.2.2 Mass transfer of components from one phase to another phase

There are two mass transfer steps in series involved for a component from the bulk of the fluid to the internal particle. The overall mass transfer coefficient k for both steps is

$$1/k = 1/k_f + 1/k_{diff}$$

The external mass transfer coefficient in the fluid k_f can be calculated from
(For fluids passing through a packed bed Ranz WE, Chem. Eng. Prog. 1952, 48, 195 p. 247.)

$$Sh = 2 + 1.8\,Re^{0.5}\,Sc^{0.33} \tag{19.1}$$

For $Re = 10$, and for gases $Sc = 2$, this results in $Sh = 10$.

The internal mass transfer coefficient governed by the internal diffusion is D_{in} and the diffusion length. The internal diffusion coefficient is at least a factor 4 lower than the external diffusion coefficient, while the diffusion length is about $0.5\,d_p$. This means that the internal diffusion is the limiting rate parameter in the same way as it is for porous catalysts treated in Chapter 10.

Design for internal mass transfer

The internal mass transfer (sometimes called adsorption kinetics) parameter k_{in} for adsorption can conservatively approximated by

$$k_{in} = a\,2\,D_{eff}/d_p \tag{19.2}$$

The specific outer surface area a for a sphere is

$$a = 6/d_p$$

The internal diffusion coefficient D_{eff} is a factor 4 to 10 smaller than free fluid diffusion, so the internal diffusion is still the lowest value, so the limiting mass transfer. For a concept design, the internal kinetic rate parameter k_{in} can be taken as conservative governing parameter.

The zone length where adsorption takes place cannot be approximated by taking $k_{in}t = 3$, which is equivalent to 95 % conversion of fluid component to adsorbed component.

The length of the mass transfer zone is L_{MTZ} is then approximated by

$$L_{MTZ} = v_{sup}/k_{in} \qquad (19.3)$$

The actual bed length can now be chosen. It has to be larger than $L_{\backslash MTZ}$; the transition zone. Beyond which the value of the bed length means an even higher efficiency of using the adsorption material. This is the reason that adsorption beds are in general chosen to be long. Pressure drop limitation is in general the limiting factor. The simplest method of designing an adsorption packed bed is provided by de Haan [1]. Albright [2] provides a more elaborate design method.

19.2.3 Residence time distribution of fluid

To avoid negative aspects of residence time distribution for the adsorption bed design the residence time distribution of the fluid flow should be so narrow that the adsorption flow can be treated as plug flow. The Peclet number approach is suitable for this design criterion. See for details of this reasoning in Chapter 8 on reactor design. Plug flow for the gas is obtained if the actual Peclet number is larger than the minimum Peclet number Pe_{min}.

$$Pe_{min} = (100/\text{error percentage})\, n\, \ln(1/(1 - X)) \qquad (19.4)$$

For a 1 % error percentage and assuming the adsorption can be treated as a first-order reaction ($n = 1$) and for a conversion of $X = 0.90$. This results in $Pe_{min} = 300$.

Choosing $Re = 10$ results in $Pe = 2(L/d_p)$.

This results in a minimum bed length for plug flow behavior L_{min}. $L_{min} = 150d_p$.

Take for example a particle size of 1 mm The bed length must then be longer than 0.15 m, to avoid negative residence time distribution effects.

By choosing $Re = 10$ value, a minimum superficial fluid velocity v_{sup} value can now be obtained.

With that also a maximum bed diameter can be determined, just by using the required design feed flow capacity.

19.2.4 Heat effects and heat-up

Heat effects in general play a role in adsorption. A simple way of dealing with the heat effect in adsorption is to calculate the maximum temperature rise by assuming that all heat of absorption is used to heat up the fluid. Hence, neglecting the heat up capacity of the adsorption bed. This then results in the maximum temperature rise. The maximum temperature rise can then be estimated from

$$\Delta T_{admax} = (\Delta H_{ad} \cdot M_{ad})/F \cdot Cp_{fl} \tag{19.5}$$

ΔT_{admax}: (K) Maximum temperature rise fluid by adsorption heat
F: (kg/s) Feed flow rate
Cp_{fl}: (J/(kg K)) Specific heat capacity feed flow
ΔH_{ad}: (J/(mol)) Heat of adsorption per mol of absorbing component
M_{ad}: (mol/s) Adsorption component feed flow rate

If the maximum temperature is less than 1 °K, then the inlet temperature can be used for the whole adsorption bed. If the maximum temperature is higher then the inlet temperature plus this maximum temperature rise can be used as representative for the bed temperature and the equilibrium value K, for the cases where K decreases with increasing temperature. If K increases with temperature then the value of K for inlet temperature can be taken as a conservative estimate.

Xiao [3] provides details of thermal effects.

19.2.5 Heat transfer

Heat transfer from the fluid to the adsorption material will be rapid, due to the large specific surface area between the fluid and the adsorption particles. But for adsorption cases where the temperature rise is small heat transfer is not important for the adsorption step.

If the recovery step is by thermal swing then heat transfer is important. But that step is omitted from this design chapter for reasons given in Section 19.1.

19.2.6 Impulse transfer

Impulse transfer from the fluid to the adsorption bed can be calculated by the Ergun pressure drop correlation, provided in Section 10.2.2.3. The bed should be designed such that the pressure drop does not exceed the bulk crushing strength of the adsorption material.

19.2.7 Overall design procedure for adsorption and desorption

For an overall design, the bed diameter and bed length will be optimized. If the particle size is a degree of freedom, then small particles can be chosen at first instance, to obtain high internal diffusion rates (also called kinetics in adsorption). However, smaller particles means a higher pressure drop, so a higher compression cost. Furthermore, the bulk crushing strength of the particles should not be exceeded.

In the desorption step, the minimum fluidization velocity of the bed should not be exceeded. This means that an optimization has to take place for particle size, bed diameter, and bed length. Such a procedure is beyond the scope of this book. With an expression of Section 19.2 reasonable bed dimensions can be determined.

The number of beds also have to be chosen. The minimum number is one. First, the bed is used for adsorption. Then the feed is stopped for the desorption to take place. So, continuous operation is not feasible for a one bed design. At least two beds should be chosen. In practice, four or more beds are chosen, facilitating maximum yield of desired product on the feed. Gray (shaded) by-product streams can be minimized by smart switching.

19.2.8 Limitations of design method

Because of all phenomena described in the previous sections many design parameters are involved, each with their own uncertainty. The design will therefore have several uncertainties.

To remove these uncertainties, it is advocated to validate the concept design experimentally by a mini packed bed with the scale-down methods of packed bed reactors as described in Chapter 10.

19.3 Example case: Adsorption case carbon dioxide capture from flue gas

Overgaauw describes in detail an adsorption process design for the capture of carbon dioxide by adsorption from the flue gas of a steam methane reforming process. The feed contains carbon dioxide, nitrogen, and water, Water is first removed by a drying step. Trace amounts of NOx are assumed to be also removed prior to the adsorption [4].

The gas composition and isotherm information is provided by Haghpanah [5]. The adsorption material is activated carbon.

Four identical beds in parallel in swing operation are designed. The swing steps are: (a) adsorption, (b) forward blowdown, (c) desorption, (d) repressurization.

Table 19.1 shows parameter values and the bed size determined by the design method of Section 19.2. The table also shows the optimized bed sizes by the detailed modeling of Overgaauw [4]. The conclusion is that the boundary value for the bed length determined by method of Section 19.2 is in agreement with the design result of Overgaauw [4].

Table 19.1: Adsorption column design carbon capture case.

Parameter	Dimension	Design	Overgauw [4]	Comment
ρ_g	kg/m^3	1.3	0.9-1.6	
η_g	kg/(m.s)	$1.7\,10^{-5}$	$1.7\,10^{-5}$	
D_g	m^2/s	$1.6\,10^{-5}$	$1.6\,10^{-5}$	
D_{eff}	m^2/s	$0.25\,10^{-5}$		
v_{gsup}	m/s	0.25	0.25	Design Re = 38 of 10?
d_p	m	$2\,10^{-3}$	$2\,10^{-3}$	
L_{min}	m	0.3		
k_{in}	1/s	$1.5\,10^{-2}$		
L_{MTZ}	m	1.7		Minimal bed length required
L_{min}	m	0.3		$150 \times d_p$
L	m	>1.7	4.2	
D	m		1.2	
N	–		4	

Feed flow rate: 1,000 m^3/h and case properties obtained from Overgaauw [4].

Bibliography

[1] De Haan AB, Eral HB, Schuur B. Industrial Separation Processes: Thermal Unit Operations and Mechanical Unit Operations. Walter de Gruyter GmbH & Co KG; 2025.

[2] Albright L. Albright's Chemical Engineering Handbook. CRC Press; 2008.

[3] Xiao J, Peng Y, Bénard J, et al. Thermal effects on breakthrough curves of pressure swing adsorption for hydrogen purification. International Journal of Hydrogen Energy. 2016;41(19):8236–45.

[4] Overgaauw RP. Design and Optimization of a Carbon Capture Process. Master's thesis. University of Twente. Sourced 18 July 2025. https://essay.utwente.nl/106096/1/Rob_Overgaauw_MA_TWN_SPT.pdf.

[5] Haghpanah R, Majumder A, Nilam R, et al. Multi-objective optimization of a four-step adsorption process for post-combustion CO_2 capture via finite volume simulation. Industrial & Engineering Chemistry Research. 2013;52(11):4249–65.

20 Membrane separation

20.1 Basic membrane separation

20.1.1 Introduction to membrane separation

The field of membrane separations has its own terminology, which is very well described by Kronos [1]. We will use his terminology and make some links with chemical engineering terminology for clarification.

Membrane filtration

In membrane filtration, a component or several components are filtered from the feed stream, and stay on the feed side of the membrane. So, membrane filtration belongs to the unit operation filtration and is not further treated in this chapter.

Membrane separation

In membrane separation, the desired pure product selectively diffuses through the membrane, while all other components stay on the feed side of the membrane. This then results in a very pure product stream.

For membrane separation, several other terms are used and are explained here using Figure 20.1.

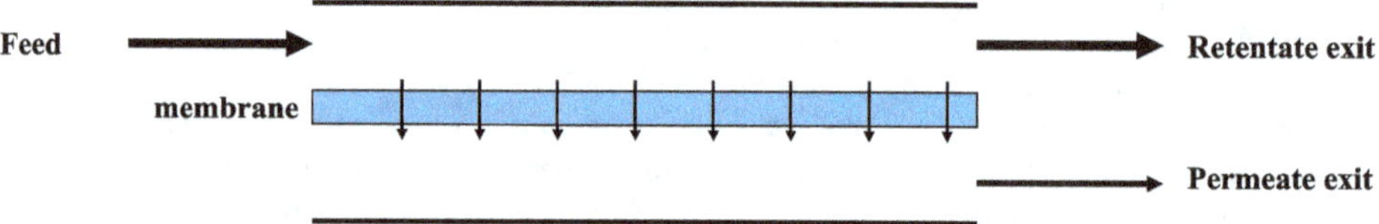

Figure 20.1: Continuous tangetial membrane separation terms.

The component diffusing through the membrane is called permeate. The mixture of all other components staying on the feed side of the membrane is called retentate. The operation can be batch or continuous.

The batch operation is called dead-end membrane separation [1]. As the feed chamber is closed at the end.

The continuous operation is often called tangential membrane separation, as the feed flows tangentially along the membrane surface through the retentate chamber and leaves at the retentate side exit.

In this chapter, only the continuously operated membrane option is treated, as the dead-end batch operation is only attractive for small capacity processes.

https://doi.org/10.1515/9783111203256-025

The driving force for membrane separation is Gibbs free energy. As the concentration on the pure product side of the membrane is higher than on the feed side, it needs a high-pressure difference over the membrane to still get a Gibbs free energy difference as a driving force.

However, a generic description of fluxes through membranes is still not available. Song, for instance, showed only recently, a correct driving force for osmotic pressure membranes, such as for producing drinking water from seawater concentration differences [2].

This chapter is limited to selective nano filtration of organic molecules in an organic mixture, where a pressure difference is the driving force, and the membrane only facilitates a selective diffusion of one molecule type through it. The transfer rate is then partly governed by viscous flow and partly by diffusion mass transfer through the membrane [3].

The permeability for the component decreases with increasing membrane layer thickness. Therefore, the actual membrane layer thickness is minimized. To keep sufficient mechanical strength, to counteract the high = pressure difference, the membrane layer is supported by a strong porous layer.

The separating driving force goes to low values on the exit side of the membrane, as the diffusion component is depleted. This means that the degree of removal of the product component from the feed stream is limited. Or in process design terms the yield of product on the feed is low. In the membrane field, the product yield is called relative recovery or substance efficiency.

For many applications, this limited relative recovery is acceptable if the feed component is available in abundance at hardly any cost. An example is the membrane separation of pure water from sea water.

For cases, however, where not only a pure product stream is required but also a high recovery of the product component from the feed stream, then a hybrid solution can also be considered in which the retentate components are separated in a second step and the second outlet containing mainly the product component is recycled back to the membrane section. This subject is treated in Section 20.2.1 and also in the design case (Section 20.3).

Process concept design around the membrane will in general be needed for a complete process design solution.

20.1.2 Background mass transfer performance

Designing a membrane on the same base as for other unit operations described in Chapters 16–19 is not feasible, as a reliable quantitative description of mass transfer with expressions for Sherwood and Peclet as functions of Reynolds is still lacking. Ververs compared nine external mass transfer correlations for tangential membrane systems.

The predictions varied wildly and none of them predicted the experimental results of his research [4].

There are at least five reasons for this lack of reliable mass transfer correlations:

Reason 1. For membrane purifications: The fluid flowing through the retentate chamber has a significant radial flow component, due to the fact that part of the fluid flows convectively through the membrane. Hence, a mass transfer correlation of Sherwood versus Reynolds based on diffusion through a boundary layer is not applicable.

Reason 2. The retentate channel has a rough surface on the membrane side. This roughness has an effect on mass transfer and should be properly included in the mass transfer correlation. However, this has not properly been done so far in mass transfer coefficient correlations.

Reason 3. Turbulence promotors and spacers are often applied to enhance external mass transfer, but no generalized expressions for their effects on mass transfer are available.

Reason 4. The residence time distribution of the flow in the retentate channel for the component to be transferred also plays a role in the membrane performance but this is also not properly taken into account in mass transfer correlations.

Reason 5. At the entrance of the feed to the retentate channel, the fluid flow pattern is developing, so there is an entrance effect. Some researchers have incorporated this effect by adding a term L/d_h in the Sherwood correlation. However, this entrance effect should be taking into account in a different way by having a mass transfer correlation for the entrance section and one for the fully developed flow section. This information is not available either in the public domain.

This means that a classical chemical engineering design approach based on mass transfer correlation of Sherwood versus Reynolds and Schmidt dimensionless numbers cannot be applied.

For this reason, an experimental approach for the mass transfer at the retentate side of the membrane, and for the transport through the membrane is presented here.

20.2 Membrane design method

20.2.1 Scouting laboratory scale experiments

Scouting experiments with a lab set-up will give an indication of the best membrane material and best temperature and pressure is to be used. The set-up can be a short membrane module typically with a length of 20 cm such as applied by Cespedes [5].

The experiments should be carried out for a variety of feed compositions including the feed composition similar to the envisaged commercial scale.

20.2.2 Empirical concept design method

To get flux information for commercial scale design purposes, the following empirical approach is proposed based on the experimental set-up of Ververs [4].

Step 1: Experiments for concept design

Design a short membrane, typically with a length of 0.5 m and with similar geometrical dimensions as Ververs described in Section 20.3.

Perform an experimental program with various fluid velocities and pressure differences over the membrane. The experiments should include a membrane performance in product yield desired for the commercial scale design.

Step 2: Commercial scale process concept design

The commercial scale membrane set-up can be designed by making the membrane a factor 5–10 longer than the experimental unit, but assuming the same flux through the membrane. The highest flux values obtained in the experimental set-up for the desired depletion of the product component on the retentate side are used as input value. Because of the longer membrane, the fluid velocity in the retentate chamber will be higher; hence, mass transfer coefficients will be higher; hence, this results in a conservative design.

The total require capacity for the commercial scale is obtained by applying multi-tubular design. A differential model for the membrane can be developed later to optimize the membrane configuration in the same way as applied by Abdul Majid [6].

The commercial scale concept design should also include measures to counteract membrane fouling. This is a big subject with many aspects. But the design should include measures to counteract fouling.

Often the fouling on the retentate side of the membrane can be removed by a rinse step. This reversible membrane fouling means that a de-fouling process step has to be incorporated in the design. This can be done as follows. The normal membrane operation is stopped and a rinse with a solvent is then applied to remove the fouling components from the membrane. The solvent rinses the membrane and results in a waste stream of solvent plus fouling component.

The dirty solvent in general may be regenerated by a distillation step to recover most of the solvent and the resulting heavy bottom stream may be incinerated.

This means that the solvent recovery step is part of the process concept design. The solvent cost and the distillation cost can then be incorporated in the cost estimate of the process concept design.

Step 3: Evaluate commercial scale design

With that, commercial scale design of step 2 evaluates the design and includes a cost estimate on investment and energy cost.

20.2.3 Limitations of design method

The presented design method is based on plausible reasoning but is not a result of scientific research. Its value is therefore limited. In the future, better design methods may be generated by membrane design research.

20.3 Example case: Organic Nano-membrane Solvent (ONS) separation case

20.3.1 Process concept design

Organic nano-membrane solvent separations are applied at large commercial scales since 1998 [3]. So, this separation is of industrial relevance.

Patrick Schmidt [3] provides an interesting design case for membrane separation. It is the alkylation of toluene with propene resulting in products iso-butyl benzene an n-butyl benzene. A large surplus of toluene over propene is used in the reaction section so that a deep conversion of propene is easily obtained.

The large surplus of toluene, however, would require a large distillation column to recover the surplus toluene for recycling back to the reactor. This would mean a large distillation column and a large amount of energy requirements, as toluene is the lighter component recovered from the distillation top as shown in Figure 20.2.

By applying an Organic Solvent Nano-membrane (OSN) separation up front of the distillation, the distillation column size and the required energy is greatly reduced.

The membrane separation involves a reasonably selectivity of butylene benzene products as retentates, and toluene as permeate. The toluene permeate flow is recycled back to the reactor. The retentate fed to a distillation column to separate toluene from the butylene benzenes.

As the selective rejection of butylene benzenes by the membrane is not perfect, two membranes are designed for by Schmidt with a line-up as shown in Figure 20.3.

We have added a bleed stream to counteract membrane fouling including a distillation step for organic liquid stream recovery as shown in Figure 20.4.

These two membrane line-ups make the design also robust to some leakage of butylene benzenes with the toluene through the membranes.

This nano-organic solvent membrane separation is a very suitable application as no deep recovery of toluene is needed. And also, a little rejection of butene products by the membranes does not mean product loss, as the retentate stream is recycled back to the reactor.

We have taken this case to illustrate the empirical design and scale-up method of Section 20.2.

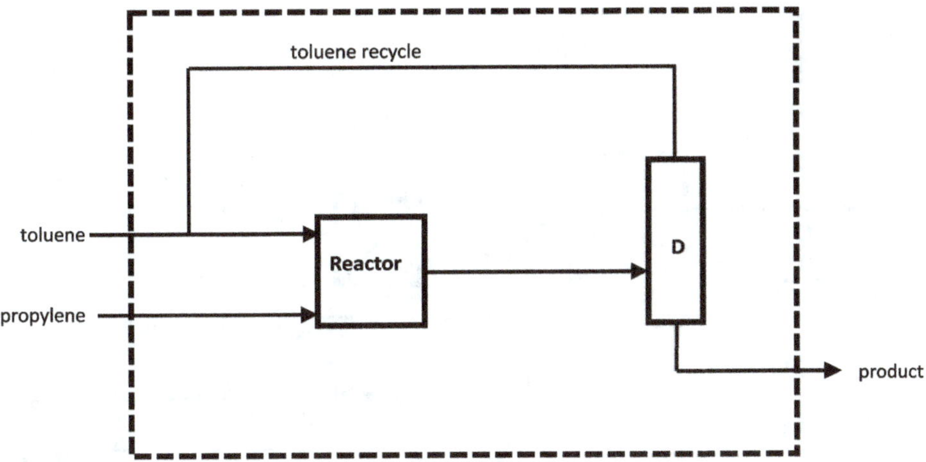

Figure 20.2: Toluene proylene alkylation process base case design.

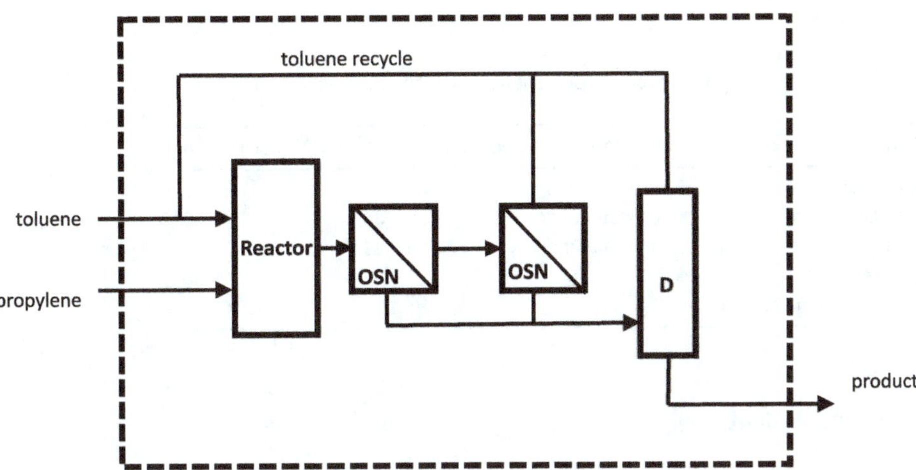

Figure 20.3: Toluene propylene alkylation process with OSN membranes.

20.3.2 Economics concept design and its development

Table 20.1 shows the investment cost per ton of product. The absolute values will not be accurate, but the 40 % cost reduction by the membrane set-up, which means 350,000 €/y is significantly high to warrant a feasibility design study. In that design study, a heat exchanger up front of the membrane may be designed for and also a membrane cleaning process step. An improved cost estimate for the commercial scale design, the cost of a development program and the incentives to warrant execution of that development program will be determined.

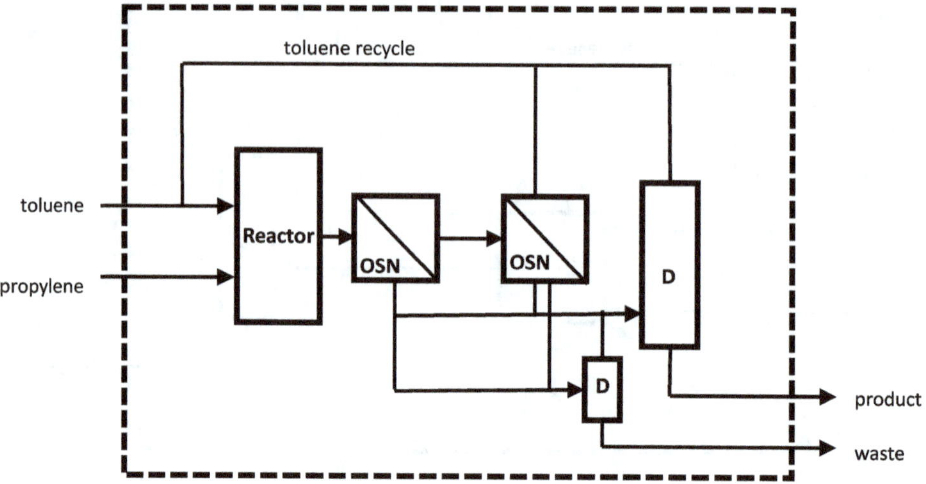

Figure 20.4: Toluene propylene alkylation process with OSN membranes including membrane fouling removal step.

Table 20.1: Cost calculations of base case design and OSN membrane design [3].

Parameter	Dimension	Base case design	OSN case design
Membrane cost	€/ton isobutene	0	9
Distillation cost	€/ton isobutene	85	40
Total	€/ton isobutene	85	50
Cost reduction	%	0	40
Cost reduction per year	€		350,000/y

20.3.3 Pilot plant design and testing for scale-up

The experimental set-up of Schmidt was a laboratory scale crossflow membrane. No details are provided on geometries. His program was to select the best membrane material, optimize for temperature and pressure, and obtain the desired selectivity.

For scale-up purposes, it is proposed to design a single tube membrane with the same membrane material and conditions as the intended commercial scale and also the same single tube dimensions and flow conditions.

The length of the pilot plant membrane tube may be selected to be shorter than the commercial scale tube. The fluid flow rate per membrane area should then be reduced by the same factor. This approach is followed here.

Table 20.2 shows the design parameters of the commercial scale and of the downscaled pilot plant. The commercial scale plant has longer tubes, which means that the linear fluid velocity through the retentate chamber is higher; hence, the external mass transfer coefficient will be higher. So, the design based on the pilot plant data will be conservative.

Table 20.2: Concept design parameters organic solvent nano-membrane separation.

Parameter	Dimension	Experimental unit	Commercial design
Isobutene + n-butene production	kg/a		$10\,10^6$
Isobutene + n-butene membrane feed	kg/s		0.3
Butylene benzenes feed fraction to membranes	%	10	10
Overall toluene selectivity membranes	%	99	99
ΔP over membrane	Bar	50	50
Membrane flux of toluene	$kg/(m^2s)$	0.017	0.017
Membrane area per membrane step			18
Membrane length	M	0.7	5
Membrane outer diameter	10^{-3} m	14	14
Membrane single tube area	m^2	0.03	0.2
Number of tubes per membrane step	–	1	90

Conditions for experimental and commercial design by Schmidt [3].
Membrane material: Puramen™ 280.
Temperature: 30–50 °C.

Bibliography

[1] Koros WJ, Ma YH, Shimidzu T. Terminology for membranes and membrane processes. Journal of Membrane Science. 1996 Nov 13;120(2):149–59.

[2] Song L, Heiranian M, Elimelech M. True driving force and characteristics of water transport in osmotic membranes. Desalination. 2021 Dec 15;520:115360.

[3] Schmidt P. OSN-assisted reaction, and distillation processes. In: Lutze P, Gorak A, editors. Reactive and Membrane-Assisted Distillation. Berlin: De Gruyter; 2016.

[4] Ververs WJR, Ongis M, Arratibel A, Di Felice L, Gallucci F. On the modeling of external mass transfer phenomena in Pd-based membrane separations. International Journal of Hydrogen Energy. 2024;71:1121–33.

[5] Cespedes S, Blankert B, Das R, Altmann T, Picioreanu C. Improved modeling and measurements of mass transfer in spacer-filled feed channels of membrane separation processes. Separation and Purification Technology. 2025 Sep 7;367:132946.

[6] Abdul Majid O, Kuznetsova M, Castel C, Favre E, Hreiz R. Impact of concentration polarization phenomena on gas separation processes with high-performance zeolite membranes: experiments vs. simulations. Membranes. 2024 Feb 1;14(2):41.

21 Mixer design

21.1 Basics mixing

21.1.1 Mixing definition

The term "mixing" is used in various ways in the field of chemical engineering. In this chapter, it is about mixing miscible fluids to obtain a homogenous fluid. Another way of defining mixing is diminishing concentration differences. Figure 21.1 shows three degrees of mixing used in chemical reaction engineering [1]. So, this chapter is about obtaining ideal micro mixing in a single-phase fluid, be it a gas or a liquid.

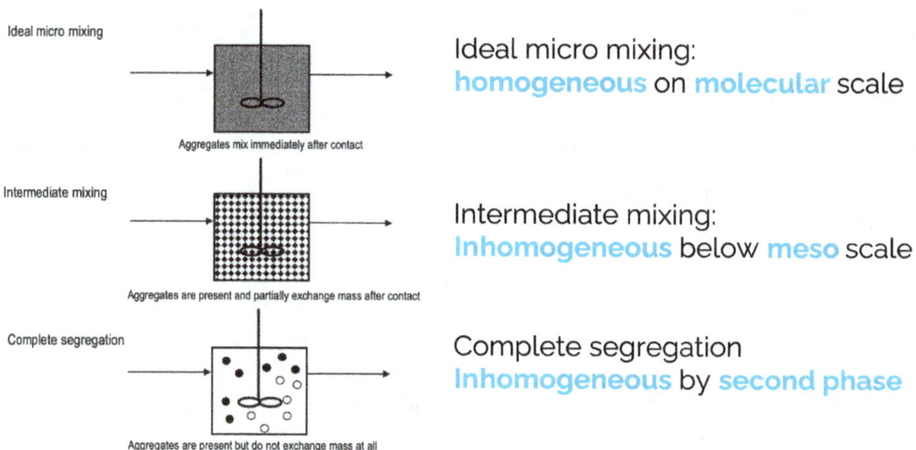

Figure 21.1: Degrees of mixing in single phase [1].

This micro mixing is needed for chemical reactions. It is also needed for many other applications, such as product formulation in food and pharma, where various ingredients have to be mixed down to molecular scale, to obtain a homogenous active product.

To achieve that the product outlet stream is completely micro mixed, a narrow residence time distribution of the fluids flowing through the mixer device will also be needed. This can be achieved in a batch mixing operation and in a continuously operated mixer with near plug flow residence time distribution behavior.

The focus in this chapter is then about selecting a suitable mixer type and about a concept design to achieve this complete micro mixing.

The knowledge field of mixing is large and complex, with many aspects. This chapter is limited to mixing fluids in turbulent fields. So, laminar mixing of highly viscous liquids is not treated.

https://doi.org/10.1515/9783111203256-026

21.1.2 Turbulent mixing theory

Turbulent fields are characterized by eddies. Figure 21.2 shows a picture of these eddies. The largest eddies, the size of the mixer configuration, whirl around with a high velocity, similar to the primary velocity of the fluid in a pipe or the pump velocity of the stirrer. These large eddies cause macro mixing, meaning that blobs of unmixed material are distributed over the whole fluid. These large eddies create smaller and smaller eddies until the smallest eddy size is reached. At this small size, the inertia energy is equal to the viscous shear energy. These smallest eddies are called Kolmogorov eddies, because Kolmogorov derived an expression for the size of these eddies [2].

Concentration differences diminish inside these small eddies only by diffusion. The time needed to obtain near uniform concentrations inside these eddies is called micro mixing time. This last step is the slowest step and also an essential step for completing the mixing. So, the mixing time by diffusion of the smallest eddy is governing the micro mixing time.

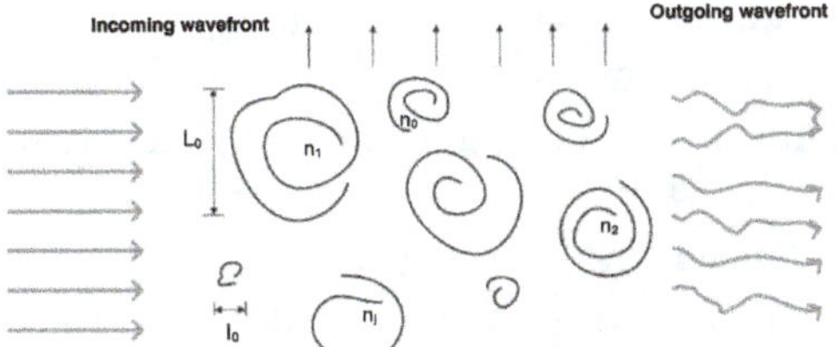

Figure 21.2: Turbulent mixing from large eddies to smallest eddies.

Corrsin proposed to use the Fourier number (Fo) to calculate the micro mixing time of the Kolmogorov eddy [3]. He proposed to choose a value Fo = 0.08 to calculate the micro mixing time for the Kolmogorov eddy size. Figure 23.1 of Chapter 23 shows that then concentration in the center of the eddy for spherical particles deviates only 1 % from equilibration value. So, this proposed value of the Fourier number makes sense.

For most applications, this degree of micro mixing is sufficient. If, however, a lower concentration difference than 1 % deviation is needed, then Figure 23.1 can be used to find a higher Fourier number value. If, for instance, if a 0.1 % deviation is needed, then the Fourier number value should increase by a factor 1.6 to Fo = 0.13.

Here follows a quantitative expression for the micro mixing time

The time required for equilibrating concentrations in the smallest eddy by diffusion to a standard deviation of 1 % can be calculated from equation (21.1):

$$t_m = l_k^2/12\,D \tag{21.1}$$

t_m = mixing time of Kolmogorov eddy (s)
D = diffusion coefficient (m^2s^{-1})

The Kolmogorov eddy size l_k is given by

$$l_k = \left(\vartheta^3/\varepsilon\right)^{1/4} \quad \text{(m)} \tag{21.2}$$

in which
ϑ is the kinematic viscosity of the fluid ($m^2\,s^{-1}$)
ε is the energy dissipation ($W\,kg^{-1}$)

By inserting l_k into expression (21.1) results in the micro-mixing time expression (21.3)

$$t_{micro} = 17(\gamma/\varepsilon)^{1/2} \quad \text{(s)} \tag{21.3}$$

So, the micro-mixing time for turbulent fields only requires the kinematic viscosity of the fluid and the power dissipation.

To get an impression of the time required for fluid to be micro-mixed, let us take mixing of two aqueous solutions. The kinematic viscosity of water at ambient temperature; $\gamma = 10^{-6}\,m^2/s$. Let us furthermore assume a power dissipation; $\varepsilon = 1\,W/kg$. This results in a Kolmogorov eddy size $l_k = 3\,10^{-5}\,m$, and a micro-mixing time, $t_m = 0.1\,s$.

The kinematic viscosity of liquids can vary enormously between $10^{-6}\,m^2/s$ (water) and $1\,m^2/s$ (concentrated sugar syrup). So, the micro mixing time is strongly governed by fluid viscosity.

For cases where one of the fluids has a much higher viscosity than the other fluid, the Kolmogorov theory does not apply anymore. For cases where the ratio of fluids to be mixed is $\gg 1$, the fluid of the highest viscosity should be taken to determine the Reynolds number value and the micro-mixing time. Experimental validation will be needed to obtain a reliable commercial scale design.

The Kolmogorov micro-mixing time for turbulent fields can be used for all kinds of mixers, such as tubes, static mixers, and mechanically stirred tanks.

21.1.3 Residence time distribution effects on micro mixing

For micro mixing to be effective, all parts of the fluids to the mixer should be micro mixed when leaving the mixer. This means that the residence time distribution of the fluids in the mixer is also an important phenomenon. If the residence time distribution of the fluids through the mixer are narrow, then the required residence time to achieve this micro mixing is the same as this micro-mixing time. If, however, the residence time distribution of the fluids through the mixer is wide, then some lumps of fluids will leave the mixer unmixed. So, to achieve complete micro mixing a narrow residence time distribution of the fluids in the mixer is required.

21.2 Mixer selection and design method

21.2.1 Mixer selection

Main equipment options for turbulent mixing of fluids are:
- Tube mixers
- Static mixers
- Mechanically stirred tank mixers

Table 21.1 shows the main characteristics of these mixers.

Table 21.1: Turbulent mixer characteristics.

Mixer	Turbulent field commercial scale	Residence time distribution	Applications
Tube	Predictable	Narrow for $L/D \gg 1$	Low viscosity and density differences
Static mixer	Predictable	Narrow	Most cases
Mechanical stirrer	Not predictable	Wide	

For the mixing of two gases, or two miscible liquids, with small differences in density and viscosity; the simplest solution is mixing in a tube. The tube dimensions can be easily calculated from the correlation provided in Section 21.1.

When the viscosities and densities differences of the two liquids to be mixed are large, then the fluid flow turbulence in a tube becomes complex, stratification because of density differences may occur and the Kolmogorov theory can no longer be used.

For fluids mixing with large differences in density and viscosity a static mixer is an attractive option. Because there the turbulence is governed by the size of static mixer elements. High power dissipation can be obtained; hence, short micro-mixing times. Moreover, the residence time distribution of the static mixer is narrow and predictable.

Reliable scale-up is obtainable by experimental validation of the mixing design by a small diameter static mixer test unit.

A mechanically stirred tank mixer is the third option. It has two disadvantages when compared to the static mixer.
A) The turbulent field in inhomogeneous and this inhomogeneity increases with scale-up to commercial scale. And experimental validation of commercial design is not practically feasible.
B) For continuous operation, the residence time distribution is wide.

The turbulent field and its mixing in a mechanically stirred tank (vessel) at large scale are still not predictable. Research is still ongoing; see, for example, the PhD thesis of Mirfasihi in 2023 [4].

21.2.2 Mixing in a tube by power dissipation

For mixing of two gases, or two miscible liquids, with small differences in density and viscosity, the simplest solution is mixing in a tube. The great G. I. Taylor [5] has treated this subject in detail. He showed theoretically and experimentally that power dissipation is the dominant factor in mixing in tubes and that thereby wall roughness is an important parameter as wall roughness has a large effect on the pressure drop, and thereby on the power dissipation, resulting in more intense turbulence with a smaller Kolmogorov eddy size; hence, a shorter micro mix time.

For tube flow the power dissipation per unit mass is given by

$$\varepsilon = (v/\rho)(dP/dL) \tag{21.4}$$

With this Kolmogorov length scale l_k can be calculated and the micro-mixing time.

The tube length required for micro mixing is then

$$L_{mix} = v \cdot t_{mix} \tag{21.5}$$

The pressure drop over a cylindrical pipe is

$$\Delta P = Fa(L/D)\frac{1}{2}\rho v^2$$

So, the pressure drop per unit length can be determined from (21.6):

$$dP/dL = Fa/D\left(\frac{1}{2}\rho v^2\right) \tag{21.6}$$

Fa is the Fanning friction factor. The value of this term depends on the Reynolds number and the rough surface of the pipe. Correlations for Fa as a function of Reynolds and the tube surface roughness are provided by all physical transport phenomena textbooks such as, for instance the one by van den Akker [6].

The residence time distribution of the tube should be narrow, so for the calculated value of L_{mix} the residence time distribution should be calculated. For turbulent flows in tubes Pe = 2 L/D gives the residence time distribution.

For micro mixing to 1 % concentration differences, Pe should be Pe = 200 (100 CISTR in series).

So, L/D = 100 is sufficient for mixing.

Green experimentally proved that the feed point of the second fluid to be mixed with the first fluid should be in the center of the tube, as there the large eddies ensure rapid distribution over the whole cross-sectional area. A feed point at the wall resulted in poor mixing, due to the boundary layer, prevailing the flow behavior locally [7].

Because the largest eddies are the size of the diameter the number of feed points at the tube inlet is not critical. A single feed point is sufficient.

For concept stage design, the tube dimensions determined in this way are sufficient for cost calculations. Surface roughness is the most critical parameter for scale-up to the commercial scale. The required surface roughness should be explicitly reported to the EPC contractor company. Otherwise, they may go for a smooth tube inner surface with a low pressure drop, which is the opposite of what is required.

21.2.3 Mixing in a static mixer by power dissipation

A static mixer, also called a motionless mixer, is a tube with inserts that create radial mixing. The power dissipation is a factor 30–1,000 times higher than for an empty tube. The inserts also create a nearly homogeneous turbulent field. Moreover, the residence time distribution of static mixers is very narrow.

Static mixers are also suitable if the viscosities of the fluids to be mixed are quite different because the turbulent field is far more homogenous and far more intense. All these factors mean that static mixers are very effective and reliable mixers for any scale and for a large variety of applications.

The micro-mixing time of static mixers is studied by Green [7]. He derived his own expressions for any degree of mixing required. Here, we apply the Kolmogorov smallest eddy theory coupled with diffusion for concentration difference reduction by diffusion inside the smallest eddy. Here are the expressions for micro-mixing time:

$$t_{micro} = 17(\gamma/\varepsilon)^{1/2} \quad (s) \tag{21.7}$$

γ is kinematic viscosity (m^2/s)
ε is power dissipation (W/kg fluid)

Power dissipation by any mixer is given by

$$\varepsilon = (dP/dL)(v/\rho) \tag{21.8}$$

(dP/dL) is pressure drop/length (N/m^2)/m
v is fluid velocity (m/s)
ρ = fluid density (kg/m^3)

Pressure drop correlations for all kinds of static mixers are available in the public domain.

Some blending time correlations for some static mixers are available from Green [7] and Montante [8].

A static mixer can also be used for laminar flow mixing. Albright provides a nice overview of mixers and the required L/D ratio to obtain a degree of mixing. For a 1 % deviation in concentration differences, typically L/D = 10 is sufficient for a SMX static mixer up to a viscosity ratio of the two liquids of 100 [9].

The scale-up method of static mixers is simple and reliable. Just keep structure size, fluid velocity, and tube length for the commercial scale the same as for the test unit. Only increase the diameter to obtain the required capacity. This means that the same degree micro mixing is obtained due to
– Homogeneous turbulence
– Same energy dissipation
– Same residence time

21.2.4 Mixing in mechanically stirred tanks by power dissipation

Mixing in mechanically stirred tanks occurs as follows. The impellor creates at the impeller blade a high shear, which in turn creates a highly turbulent field locally. Further away from the impeller the turbulent eddies decay into a lower turbulent intensity. Micro mixing occurs at the smallest eddy.

At larger scale, this difference in turbulence intensity increases in an unpredictable way. By this phenomenon, reliable scale-up designs for stirred tanks are hard to make. A good introduction into turbulence in mechanically stirred tanks is a study by Professor van den Akker [9].

The pharmaceutical and fine chemicals industry often use these mechanically stirred tank reactors for mixing and other process functions in batch operation. The uncertainty of the scale-up effects is solved by having several pilot plants; increasing in size, often at a scales typically of 0.001, 0.01, 0.1, and $1\,\mathrm{m}^3$. The recipe and the batch treating time are modified at each scale-up step to combat at each scale the observed performance deviations, so that finally a reliable production at the large scale is obtained.

Limitations of stirred tanks

Mixing in stirred tanks has the following limitations. The first limitation is that a single stirred tank for obtaining a degree of mixing with a standard deviation of 0.01 will be practically impossible, due to the large macro residence time distribution causing the majority of the fluid leaving the stirred tank unmixed, it will require an enormous oversize of the average residence time of the mixing timing. It is therefore essential that for continuous operation a large number of stirred tanks in series are applied. Full micro mixing in a single stirred tank can only be obtained by batch operation.

The second limitation is that the turbulence intensity in a stirred tank is very inhomogeneous. Near the impeller, the intensity is high. Far away from the impeller, the intensity is low. This inhomogeneity increases with scale-up in an unknown way. Hence, the only reliable scale-up method for stirred tanks is the empirical scale-up method, described in Chapter 25. This scale-up is expensive if the experimental set-up is not available but has to be installed.

21.2.5 Temperature increase by power dissipation

Due to the power supplied to the fluids, the temperature of the fluids to be mixed will increase. This increase will for most cases be small. The maximum temperature increase can be simply estimated by the energy balance of power input and the heat accumulation caused by that power input. The maximum temperature rise in the mixer is then given by

$$\Delta T_{max} = \varepsilon \cdot M/(F \cdot C_p) \qquad (21.9)$$

ΔT_{max}: (°K) Maximum temperature rise fluid
ε: (W/kg) Power dissipation
M: (kg) Total fluid mass in mixer
F: (kg/s) Total fluid feeds to mixer
C_p: (J/(kg °K)) Specific heat of fluids

The total fluid mass in the mixer M can be estimated from the total mass feed time the average residence time in the mixer, which will be the total mixing time.

21.2.6 Limitations of design method

The presented design method is limited to turbulent mixing. Laminar mixing such as with highly viscous flows is not treated.

The second limitation is that it is only applicable for cases where the Kolmogorov turbulence model is applicable. So, it is not applicable in mixers with complex geometries.

The third limitation is that physical properties have to be known, in particular, the viscosity of the fluids to be mixed.

The fourth limitation is that impulse transfer of fluids on mixer internals is not treated. For high intensity mixers, these forces can be high, and special attention has to be paid on details of their construction. This attention will have to be paid in the engineering, procurement, and construction stage by the engineering contractor.

21.3 Example case: Molasses water mixer design for a fermentable feed

21.3.1 Introduction to industrial case; Molasses dilution

A biotechnology provider company decides to install a new fermenter for the production of bioethanol from molasses. The molasses will be obtained from a local beet sugar factory.

A newly recruited chemical engineer, called Barbara, has to make a concept design for the whole process. From the marketing department, she gets information that the process capacity should be 100 kton ethanol/year.

She has to design for a homogeneous feed to the fermenter. It is a bubble column with an external circulation with a heat exchanger to control the fermenter temperature.

One of the critical-process steps is diluting the molasses feed with water by a factor 10. So, this means a mixing step. She obtains the molasses composition and viscosity from a report by Olbrich [10]. The data about the molasses composition and viscosity is given in Table 21.2.

Table 21.2: Molasses physical properties.

Content and property	Content fraction (% m)	Viscosity m^2/s 20 °C	Viscosity m^2/s 50 °C
Water	16.5		
Sugars	53.0		
Gummy	19.0		
Inorganics	11.5		
Total	100.0		
Viscosity kinematic		0.6	0.06

Barbara then finds out that she has to dilute the molasses with water by a factor 10 to make it suitable for yeast fermentation.

The question is how to dilute the molasses to a homogeneous aqueous solution. From a chemical reaction engineering course, she remembers that a homogeneous feed stream to the fermenter is essential for fermentation. She also considers that the gummy part of the molasses may cause sticking to the fermenter wall, as she knows that chewing gum sticks to anything.

Barbara wants to do her job well, so he starts reading about how to mix liquids. She also uses her imagination to produce surprising solutions. From the literature, she finds out that mixing can be done by a tube, by a mechanically stirred vessel, and by a static mixer. She learns from mixing theory that any turbulent field can do the job, given sufficient residence time. This leads her to the additional option of using the feed pump to the fermenter as a mixer in combination with the tube connecting the pump to fermenter. So, she now has four options; shown in Table 21.3.

She also has taken a problem analysis course in which he learned that all options should be evaluated against the same criteria. Hence, she asks an experienced process engineer of the company which criteria to take into account for a decision about which of the four options to select.

The experienced process engineer is surprised by the systematic approach of the young chemical engineer and starts to like her. So, he informs her elaborately about how results of concept projects in general are evaluated by company management. He mentions first two hard criteria.

First hard criterion: the concept design should be technically sound. This means that a commercial scale concept design should contain only reliable unit operations. This means for the mixer that the feed to the fermenter is completely homogeneous to the molecular level, as otherwise blobs of unmixed molasses material may stick to any internal, due to the gummy substances in the molasses.

Second hard criterion: the chosen solution should be scalable to the commercial scale application, and that it is feasible in the development stage to validate this scalability experimentally. That experimental validation may be for instance obtainable from a test facility of a technology provider of the mixer.

The experienced engineer then mentions soft criteria, investment cost, reliable production with no interruptions due to mechanical breakdown, maintenance cost should be low, electricity cost should be low, and cleaning of the mixer should be feasible by just feeding a cleaning liquid through the mixer.

Barbara puts all options and criteria in Table 21.2 and then will get all information to fill in the evaluation Table 21.3.

Table 21.3: Company criterion table for evaluating mixers.

Criterion	Pump	Tube	Mechanical mixer	Static mixer
Hard technical criteria:				
Homogeneous feed to fermenter				
Experimental validation feasible				
Soft (trade off) criteria:				
Investment cost M€				
Reliable production for a long time				
Maintenance cost				
Additional energy requirement				
Cleaning procedure easy				

21.3.2 Pump as mixer by energy dissipation

Barbara starts with calculating each mixer, the mixing time, the averaged residence time, and the residence time distribution and then draws a conclusion on achieving a homogeneous stream to the fermenter.

The power dissipation of the pump is 100,000 W. The mass content of fluid in the pump is 100 kg, so the Power dissipation is 1,000 W/kg. The kinematic viscosity of the molasses $\gamma = 0.6 \ \mathrm{m^2/s}$. The micro-mixing time expression (21.1) is applied:

$$t_{micro} = 17 \, (\gamma/\varepsilon)^{1/2}$$

The power dissipation of the pump on its content; $\varepsilon = 1,000$ W/kg. This results in a micro-mixing time of 0.3 s:

$$t_{micro} = 17 \, (0.6/1000)^{1/2} = 0.3 \, s$$

The pump capacity is $0.1 \ \mathrm{m^3/s}$. The young estimates that the pump liquid hold-up volume to be $0.1 \ \mathrm{m^3}$, so the average residence time in the pump = 1 s. This is longer than the micromixing time.

However, the residence time distribution also has to be taken into account. Barbara cannot find any information about the residence time distribution (RTD) of pumps. But given the high turbulence and the short distance between inlet and outlet, she thinks that the RTD will be wide. With this estimate, she concludes that complete micro mixing in the pump cannot be assumed. So, the pump on its own is rejected as a mixer.

21.3.3 Tube as mixer

Barbara then considers the pump and downstream tube as mixing devices. The tube dimensions are Length = 10 m. Diameter = 0.17 m. So, L/d = 60. The velocity v = 4 m/s. So, the average residence time is 2.5 s.

She assumes that the pump has already dispersed the molasses in the water stream, so that a special dispersion distributor in the tube is not needed. For the Reynold calculation, she assumes the viscosity of water, as the molasses is only 10 % of the fluid mixture. For the micro-mixing time, however, she takes the molasses viscosity.

This then results in Reynolds' number Re = 70,000. So, the flow is highly turbulent. She assumes a surface roughness of the pipe of 0.02 to obtain the value for the Fanning friction factor; Fa = −0.05. The pressure drop per unit tube length is then 200 $\mathrm{N/m^3}$. The energy dissipation is then 0.8 W/kg. Putting in the viscosity of molasses in the micro-mix time expression (21.1), results in a micro-mixing time of 0.4 seconds. So, this is much shorter than the average residence time.

The residence time distribution in the tube expressed with the Peclet number for turbulent flow in pipes Pe = 2 (L/D); Pe = 2(10/0.17) Pe = 120. Hence, the tube has a narrow residence time distribution. In conclusion, the pump plus tube could serve as a mixer.

Then she tries to answer the question of how to experimentally validate this mixer design. She asks the experienced process to engineer how to experimentally validate the pump plus tube mixer option.

He advised her to first write a document about her calculations and conclusions on the pump plus tube mixing, with her name as author and with the writing date. This is for the time being sufficient proof that she is the inventor of this design.

He advises second to contact the Contract Research Organization (CRO) of their country; to discuss whether the CRO can and will do pump plus tube experiments for mixing molasses with water.

She does so and indeed the CRO is willing to perform the experiments in principle. The CRO stresses that a contract will be needed and that the CRO may claim to generalize the results and sell a license to other companies. Barbara then says that it is her idea and that she wants her name on the patent and that her company will be given a free license. She writes minutes of meeting will all conclusions and statements of the meeting and sends them to the CRO organization. She then reports internally that experimental validation of mixing of water and molasses in a pump plus tube is conceivable by going to the CRO.

21.3.4 Mechanical mixer

Barbara then considers to put a mechanical mixer tank up front of the fermenter.

She starts with assuming a power dissipation ε of 10 W/kg.

The micro-mixing time is then

$$t_{micro} = 17 \, (\gamma/\varepsilon)^{1/2}$$

$$t_{micro} = 17 \, (0.6/10)^{1/2} = 3 \, s$$

Because of the back-mixed behavior of this mechanical mixer tank, she assumes that the average residence time needs to be a factor 100 larger to get less than 1 % unmixed fluid. So, the averaged residence time needs to be 300 s. With water feed plus molasses feed of 0.1 m³/s, this results in a tank volume of 30 m³.

She then wants a ballpark capital investment figure for the mechanical mixer tank. Chapter 5 provides a correlation for the capital investment for a process step. For process capacities >60 kton/year, the capital investment cost correlation with inflation factor correction of 5.4 % for 2025 from 2024.

The capital cost correlation is then

$$C_{DPC} = 23 \times 10^3 \times N \times (Q/S)^{0.675}$$

The annual capacity of the fermentation process is Q = 100,000 t/a.

The process has no recycling. So, S = 1. She takes the mechanical mixer tank as one process step, so N = 1.

$$C_{DPC} = 23 \times 10^3 \times 1 \times (100{,}000)^{0.675}$$

$$C_{DPC} = 5\,10^6\ \$$$

21.3.5 Static mixer

It is clear from static mixer descriptions found in the open literature that mixing molasses with water in a static mixer is feasible. She assumes a power dissipation ε of 100 W/kg; this is a factor 10 larger than the mechanical stirred tank. This results in a micro-mixing time $t_{micro} = 1.3$ s.

The residence time distribution of the static mixer will be narrow, so the residence time of the mixer is set at 1.3.

She then wants a ballpark capital investment figure for the static mixer. She uses the same ballpark figure of a process step as in Section 21.3.4. So, the capital investment = $5\,10^6$ \$.

21.3.6 Mixer option evaluation

Barbara then puts all information into the evaluation Table 21.4.

Table 21.4: Mixer-option evaluation results for molasses water dilution.

Criterion	Pump	Tube	Mechanical mixer	Static mixer
Hard criteria:				
Homogeneous feed to fermenter	no	yes	no	yes
Experimental validation feasible	no	yes	no	yes
Soft (trade-off criteria):				
Investment cost M€	0	0	5	5
Reliable production for a long time	yes	yes	no	yes
Maintenance cost	low	low	high	medium
Additional energy requirement	0	0	yes	yes
Cleaning procedure feasible	yes	yes	unknown	yes

She concludes from the hard criteria results that the pump and the mechanical mixer are not valid options as the commercial scale design cannot be reliably experimentally validated at a smaller scale.

The tube (plus pump) option and the static mixer commercial scale design option can be experimentally validated as the mixing time correlation with pressure drop and Reynolds can be validated at a small test scale.

The cost of the experimental validation of pump plus tube is unknown at that moment in time. She will have to go back to the CRO to get a cost estimate.

The cost of the static mixer is also unknown. So, she will need to go to a technology provider to get a cost estimate.

Then she reads the section on limitations of designing mixers. She wonders if she should include a heat up effect by the power dissipation of the mixers. Rather than performing a calculation, she walks to the largest pump in the existing process and feels with her hand at the inlet tube and the outlet tube of a pump. To her surprise, she feels a small temperature increase at the pump outlet. She concludes that in general the heat up by the pump and the other mixing equipment options with similar power dissipation will be small and can be neglected for the time being.

After that, she decides to go first to several technology providers to get a quote for a capital estimate for the static mixer.

One of the technology providers offers to do experimental runs in their pilot plant both for the static mixer and the pump and tube for free, if they get the contract to do the engineering, procurement, and construction of the molasses tank, the water supply, the pump, and the static mixer.

With that information, she presents concept designs of the pump plus tube and static mixer to management, who are pleased with her concept design and provide the required budget for her to go ahead with the feasibility stage.

Bibliography

[1] Harmsen J, Bos R. Multiphase Reactors: Reaction Engineering Concepts, Selection, and Industrial Applications. De Gruyter; 2023.

[2] Kolmogorov AN. Local structure of turbulence in an incompressible fluid at very high Reynolds numbers. Doklady Akademii Nauk SSSR. 1941;31:99–101.

[3] Corrsin S. The isotropic turbulent mixer: Part II. Arbitrary Schmidt number. AIChE Journal. 1964;10:870.

[4] Mirfasihi SS. Experimental and numerical investigation of mixing miscible liquids with high viscosity contrasts in turbulently stirred vessels. Doctoral dissertation. United Kingdom: The University of Manchester; 2023.

[5] Taylor GI. The dispersion of matter in turbulent flow through a pipe. Proceedings of the Royal Society of London Series A Mathematical and Physical Sciences. 1954 May 20;223(1155):446–68.

[6] Van den Akker H, Mudde RF. Mass Momentum and Energy Transport Phenomena: A Consistent Balances Approach. Walter de Gruyter GmbH & Co KG; 2023.

[7] Green A. In: Stankiewicz A, Moulijn J, editors. Re-Engineering the Chemical Processing Plant: Process Intensification. New York: Marcel Dekker; 2004. pp. 227–60.

[8] Montante G, Coroneo M, Paglianti A. Blending of miscible liquids with different densities and viscosities in static mixers. Chemical Engineering Science. 2016 Feb 17;141:250–60.

[9] Van den Akker HE. The details of turbulent mixing process and their simulation. Advances in Chemical Engineering. 2006 Jan 1;31:151–229.

[10] Olbrich H. The Molasses. Biotechnologie-Kempe GmbH. 2006.

22 Heat exchanger design

22.1 Basics heat exchangers

Heat exchangers are applied in nearly every process design to heat up and to cool down fluids and solids [1]. In the 1970s of the last century, concerns of enhanced global warming of the earth atmosphere by carbon dioxide emissions led to process heat integration measures. Special heat integration methods were developed to maximize heat recovery by placing heat exchangers in the process design. Chapter 4 of this book describes such method for heat integration.

Also, heat exchangers were optimized. For instance, special heat exchangers, based on plastic, were developed for cooling of flue gases below dew points. Also, heat exchanger baffle designs were improved by CFD modeling, preventing short cutting and obtaining narrow residence time distributions of the fluids. Nowadays Engineering, Procurement, Construction (EPC) companies have special methods for heat exchanger design.

Detailed heat exchanger-type selection and detailed engineering design will therefore in general be carried out in the Engineering Procurement and Construction (EPC) stage by a contractor company. Then heat exchanger types and construction materials will be selected using criteria for corrosion, fouling, and cleaning.

For process concept design, a preselection on the type of heat exchange concept and a simple design to estimate the heat transfer area for a crude cost estimate is in most cases sufficient. Therefore, this chapter concentrates on heat exchange principle selection and preliminary sizing of the heat transfer area.

22.2 Heat exchanger design method

22.2.1 Heat exchanger concept selection

Heat exchange is an important part of most process designs. It may be required to heat up a stream to the required temperature for reaction. It may be required to cool down a product stream to ambient temperature so that it can be stored. It may be applied to withdraw or add heat to reactor, to balance the reaction heat, and control the reactor temperature.

Heat change of a stream can be obtained by

- a heat exchanger;
- contacting a hot stream directly with a cold stream, or vice versa;
- evaporating (part of) the stream, followed by condensing the vapor by a heat exchanger;
- microwave heating.

A heat exchanger is the first option to consider.

https://doi.org/10.1515/9783111203256-027

If, however, varied rapid cooling of stream is required, if for instance the product formed in the reactor quickly reacts to undesired by-products, then rapid cooling, also called quenching, is needed. If the stream leaving the reactor is a gas, then cold liquid drops, or cold solid particles can be used for rapid cooling. In the BTG-BTL biofuels process, these two options are applied at commercial scale. The biomass is rapidly heated by fluidized particles, causing flash pyrolysis [2].

If rapid fouling of a heat exchanger is expected then droplets, or solids, for heating or cooling can be considered. An example of such cooling is in the BTG-BTL process where the hot biofuel gas is rapidly quenched in a spray tower by contacting cold biofuel liquid droplets in a spray tower. In this way, undesired consecutive reactions of the product are avoided and also fouling of a heat exchanger is avoided [2].

Rapid cooling of a liquid can also be obtained by a rapid pressure drop via a nozzle, so that flash evaporation occurs. The heat of evaporation cools the stream.

On the cold side of a heat exchanger, a boiling medium such as boiling water can be applied so that amount of cooling medium is kept small and steam is directly generated for heating duties elsewhere in the process.

Rapid heating can also be obtained by microwave heating. This applied at commercial scale for food drying, where a short residence time is needed to avoid food deterioration [3].

22.2.2 Heat exchanger size design

In process concept design, the heat exchanger can be sized for the Heat Exchange Area (HTA) by a simple expression [1]:

$$HTA = Q/(LMTD \cdot \alpha) \tag{22.1}$$

HTA: Heat Transfer Area; m^2
Q: the heat flow to be exchanged; W
LMTD: Logarithmic Mean Temperature Difference between the two streams; K
α: Heat transfer coefficient; $W/(m^2\,K)$

Figure 22.1 shows a heat exchanger in countercurrent flow with in and out temperatures of the two streams.

The LMTD for countercurrent flows is given by [1]:

$$LMTD = \frac{(T_{hin} - T_{cout}) - (T_{hout} - T_{cin})}{\ln\left[\frac{(T_{hin}-T_{cout})}{(T_{hout}-T_{cin})}\right]} \tag{22.2}$$

in which ln is the natural logarithm of x; $\ln x = {}^e\log x$.

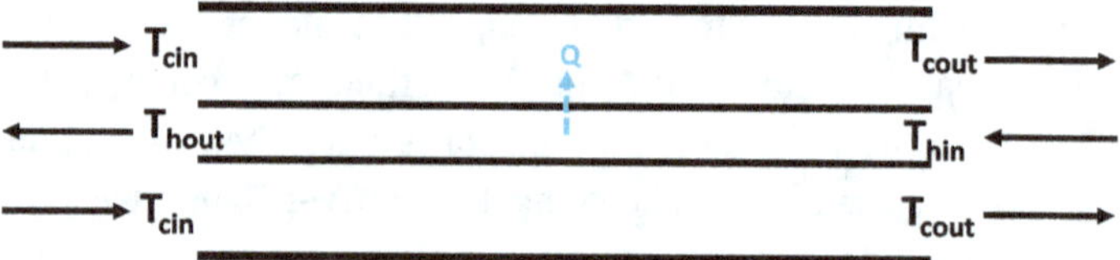

Figure 22.1: Heat exchanger concept design input parameters.

The heat transfer coefficient α is the overall heat transfer coefficient, so it lumps the heat transfer coefficients in the two fluids and the heat transfer coefficient of the heat exchange wall. These can be seen as resistances in series. This means that the overall heat exchange α can be expressed by

$$1/\alpha = 1/\alpha_{cold} + 1/\alpha_{hot} + 1/\alpha_{wall} \tag{22.3}$$

α_{cold}: The heat transfer coefficient of the cold fluid; m
α_{hot}: The heat transfer coefficient of the hot fluid; m
α_{wall}: The heat transfer coefficient of the wall

The heat transfer coefficient for the wall α_{hot} is simply λ_{wall}/d_{wall}.

In most cases, this latter wall heat transfer coefficient is much higher than the other heat transfer coefficients, and can be neglected.

The heat transfer coefficients of the fluids can be estimated with a Nusselt correlation.

For turbulent flow, the Nusselt correlation is given by [4]

$$Nu = 0.027\,Re^{0.8}\,Pr^{0.33} \tag{22.4}$$

Nu: $\alpha d/\lambda$
Re: Reynolds number; $Re = \rho v d/\eta$
Pr: Prandtl number $C_p \eta/\lambda$

For water streams through the heat exchanger with a velocity of 4 m/s Re = 4,000, Pr = 6, this results in Nusselt is Nu = 40 and α = 2,400 W/(m^2 K).

If we neglect the wall heat transfer coefficient, this results in the overall heat transfer coefficient

$$\alpha = 1,200\ W/(m^2\,K).$$

For liquids, a ballpark figure of heat exchanger is: α = 1,000 W/(m^2 K).

If liquid viscosities are far higher than that of water, then that will result in much lower α values.

For gases, a ballpark figure is α = 40 W/(m^2 K).

With this information, a guestimate of heat transfer area required can be determined.

22.2.3 Limitations of the design method

22.2.3.1 Physical transport properties

There are many physical properties involved in heat exchanger design. So, reliable property values are therefore important.

If there, large molecules are in the fluid then the viscosity property value may become uncertain and experimental determination of the viscosity for the fluid for the relevant temperature range is advised.

If liquid-solid slurries are involved, then the uncertainties of the heat exchange correlation for slurries should be taken into account, and even more so, if the solids are not hard by soft such as in biomass components.

22.2.3.2 Residence time distribution of fluids

The presented design method is based on the assumption that both fluids flow in plug flow through the heat exchanger. For shell and tube exchanger, this is often not the case in reality for the shell side of the heat exchanger. Often existing heat exchangers have been designed in the past with baffles in the shell side for which it was assumed that they would prevent shorting cutting off some of the fluid. In reality, this shortcutting still occurred.

Present heat exchanger designs have special shell side internals based on Computational Fluid Dynamic (CFD), which have been validated experimentally. These designs should be applied in the engineering, procurement, and construction stage by the EPC contractor.

22.2.3.3 Heat effects by impulse transfer

Friction of the fluid flows will heat up the fluids. The effect is in general small and can be accounted for in accurate sizing the heat exchangers in the EPC stage.

22.2.3.4 Fouling effects

In the presented design method, no attention is paid to fouling of the heat exchanger. In practice, this nearly always happens and for designs with unknown media, fouling rates will be needed determined from experiments for a reliable sizing of the heat exchangers.

22.3 Example case: Design biomass to liquid fuel

Thet BTG-BTL liquid biofuels process design is applied several times at commercial scale [2]. The process is presented in Figure 22.2. It has three process steps with heat exchange by direct contact between streams. These are:
– Hot sand contacting with biomass in a fluid bed with a rotating cone;
– Hot biofuel vapor contacting with cold liquid biofuel droplets in a spray tower;
– Superheated steam contacting with wet biomass.

Part of its success is the high yield of liquid product on biomass. That is caused by the rapid heating up of the biomass by mixing it with hot sand in a rotating cone fluid bed reactor, resulting in short residence time at high temperature. And also, by the rapid quenching of the product vapor by cold liquid droplets in a spray dryer, so that little consecutive reactions to undesired by-products are prevented. A third part of the success is the heat integration by using steam generated by the process to dry the biomass feed [2].

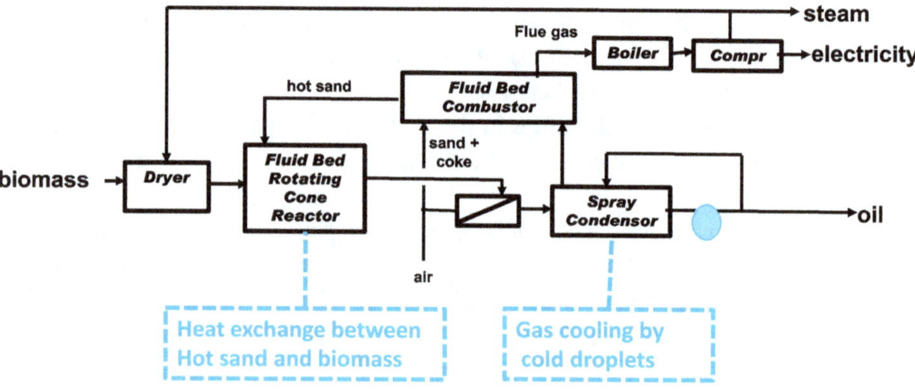

Figure 22.2: Biofuel process design BTG-BTL direct contact heat exchange.

Bibliography

[1] Seider WD, Seader JD, Lewin DR, Widagdo S. Product and process design principles: Synthesis. Analysis, and Evaluation. New York: Wiley; 2004.
[2] Harmsen J, Process VM. Intensification: Breakthrough in Design, Industrial Innovation Practices, and Education. Walter de Gruyter GmbH & Co KG; 2020 Jul 20.
[3] Stankiewicz A, Van Gerven T, Stefanidis G. The Fundamentals of Process Intensification. John Wiley & Sons; 2019. Sep 16.
[4] Van den Akker H, Mudde RF. Mass, Momentum and Energy Transport Phenomena: A Consistent Balances Approach. Walter de Gruyter GmbH & Co KG; 2023 Sep 18.

23 Solid particle design

23.1 Basics solid particle design

Solid particle processes are needed in many process industries such as food, detergents, and catalysts. The base particles may be impregnated with components from a gas or liquid phase. Often it is required that the impregnation be uniform throughout the porous particle.

The shape and size of the particle have to be designed to reach the required product performance of its application. For this particle design, the same phenomena as for unit operations play a role.

Mass transfer is governed by the mass diffusion coefficient D. Heat diffusion is governed by the thermal diffusion coefficient D_{th}. Both diffusion coefficients have the same dimension, m^2/s. The mathematical differential equation for both diffusion phenomena is the same using the dimensionless Fournier number. The solution to the differential equation is then only dependent on the value of the Fourier number Fo.

With this Fourier number, the time required for heating up a particle to a certain degree, or the time required for diffusion of a component inside a porous particle to a certain degree can be simply calculated from the Fournier number.

For gases, typical values of the mass transfer diffusion coefficient is 10^{-5} m/s. For liquids, the mass diffusion coefficient is typically 10^{-9}. So, mass transfer for gases is in general much higher than for liquids.

Section 23.2 describes all phenomena from a particle design point of view.

23.2 Solid particle design

23.2.1 Thermodynamic equilibria and properties

Thermodynamic equilibria and thermodynamic properties for solid particle impregnation are in general hard to find. Experiments will therefore often be needed to obtain this information.

23.2.2 Solid particle design for mass transfer by diffusion

Mass transfer by diffusion is governed by the same phenomenon as heat transfer by conduction. This means that for mass transfer by diffusion the same Fourier number solution of Section 23.2.1 can be used, just by replacing temperature by concentration and by replacing D_{th} by the diffusion coefficient D. The solution provided in Figure 23.1 can be used by replacing $(T_c - T_0)/(T_1 - T_0)$ by $(C_c - C_0)/(C_1 - C_0)$.

https://doi.org/10.1515/9783111203256-028

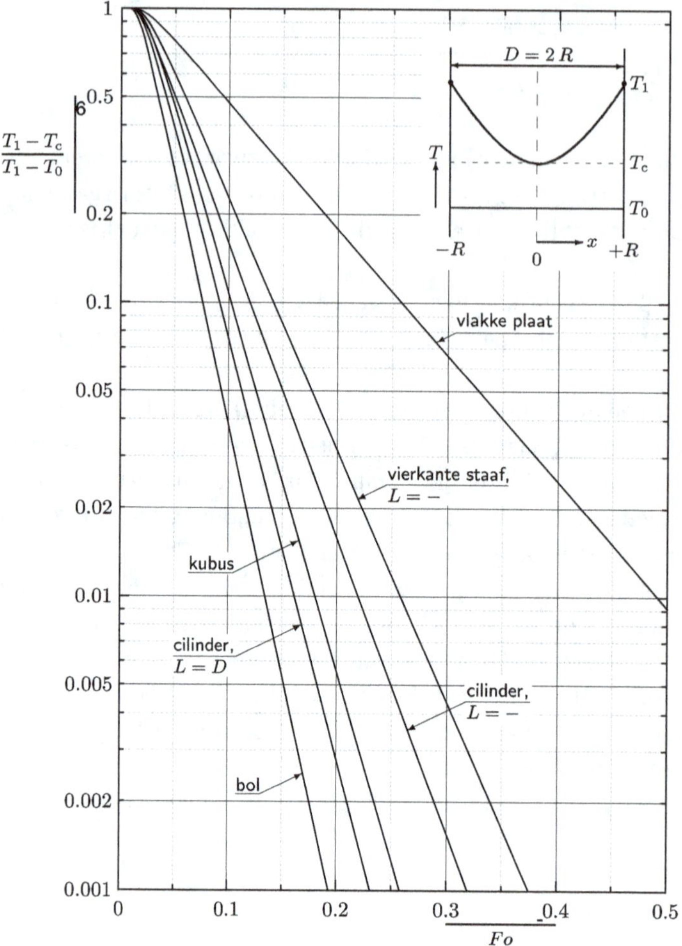

Figure 23.1: Center temperature in particles by conduction versus Fourier number [1].

The time needed for concentration equilibration for a sphere to 99 % by diffusion is then given by expression (23.5):

$$t_{equilibrium} = 0.1\, d^2/D \tag{23.1}$$

For quick estimates of diffusion times for near equilibrium, the value of the diffusion coefficient of gases is 10^{-5} m^2/s and for liquids 10^{-9} m^2/s. So, diffusion of a component in a liquid is a factor 10^4 slower than for a gaseous component.

The treatment time for particles can now be determined by expression (23.1).

23.2.3 Heat effects and heat transfer

Impregnation of porous particles by a substance can have a heat effect in the same way as for adsorption. The effect can be crudely estimated as the heat released from liquid phase to solid phase. But the real value can be higher. Dedicated experiments will be needed to determine the adsorption heat. The time needed to transfer this heat to the fluid surrounding the particle can be estimated by the method described below.

Heat conduction inside a solid particle is governed by the outside temperature of the particle T_1. The starting temperature of the particle T_0. The heating up (or cooling down) time t, the thickness of the particle d, and the thermal diffusivity D_{th}. This latter parameter has the same dimension as the molecular diffusion (m^2/s) [1]. The thermal diffusivity is given by

$$D_{th} = \lambda/\rho C_P$$

λ is the heat conductivity of the solid material.
ρ is the density of the solid material.
C_P is the specific heat of the solid material.

The time needed to heat up (or cool down) the particle is then determined by the value of the Fourier number Fo only:

$$Fo = D_{th}\, t/d^2 \tag{23.2}$$

The temperature at the center T_c can now be obtained from the dimensionless term

$$(T_c - T_0)/(T_1 - T_0) \tag{23.3}$$

as a function of the Fourier number.

For long term heating (Fo > 0.1) not only is the outer side of the particle heated up, but also the middle of the particle is heated up.

The general approximate solution with sufficient engineering accuracy is given by

$$(T_c - T_0)/(T_1 - T_0) = 1 - (4/\pi)\exp(-\pi^2 Fo) \tag{23.4}$$

[1].

So, for a desired degree of heating up of the center, the time needed can now be calculated with this expression.

Figure 23.1 shows the solution for various particle shapes.

For Fo = 0.1, for spheres, the value of $(T_c - T_0)/(T_1 - T_0) = 0.01$. So, for Fo = 0.1; heating up is completed for 99 %.

The time required for heating up a solid spherical particle can now be calculated directly from Fo = 0.1 by expression (23.4):

$$t_{heating\ up} = 0.1\,d^2/D_{th} \tag{23.5}$$

The value of D_{th} varies enormously for solid materials. Some materials have good insulation properties for a low value of λ. Others have high values of λ, and are good heat conductors. Hence, heating up times for solid particles vary enormously for different materials.

The particle diameter can now directly be designed for various particle shapes using Figure 23.1 and the Fourier number value.

23.2.4 Limitations of solids particle design

The main limitation to the presented design method is that the unit operation in which the particle is treated is not described. There are many ways of treating solid particles in unit operations, but describing these is beyond the scope of this book.

23.3 Example case: Porous particle impregnation design

A porous material needs to be impregnated with a catalytically active component by putting the particles in a liquid containing the component. The spherical particles have a diameter of 1 mm. The chemical engineer calculates that the time needed for the component to reach a near maximum concentration in the center of the particle using the Fourier number value of 0.1. Het assumes that the effective diffusion coefficient is a factor 4–10 lower than the diffusion coefficient. So, he takes $D = 1\text{--}4\,10^{-10}\ m^2/s$. The estimated times are then 250—1,000 seconds.

He advises then to perform three impregnation experiments with diffusion times for sample A = 250, B = 1,000, and C = 2,000 seconds, and afterwards test the three impregnated catalysts samples for their activity. It appears that the three impregnated catalysts show small differences in activity. The sample has a significantly lower activity than samples B and C. The latter two have similar activities, with a difference of a few percent. An impregnation time of 1,000 seconds is then chosen.

Later on, the client of the catalyst firm wants the catalyst particles to be 2 mm, to overcome pressure drop limitations. The time for impregnation and catalyst activity measurements is not available. The chemical engineer then advises to take factor 4 longer impregnation time, so 4,000 seconds.

Bibliography

[1] Van den Akker H, Mudde RF. Mass, Momentum and Energy Transport Phenomena: A Consistent Balances Approach. Walter de Gruyter GmbH & Co KG; 2023 Sep 18.

24 Integral process scale-up

24.1 Basics integral process scale-up

Process scale-up by design is a unique essential element of chemical engineering. Due to the development of process unit operation design methods, pilot plant designs and commercial scale designs could be executed with the same design method, by which scale-up to commercial scale was speeded up and also became more reliable.

However, process scale-up to commercial scale production still often goes wrong. Independent Project Analysis (IPA) revealed that about 50 % of all new commercial scale processes fail [1]. These failures, however, can be prevented by taking established precaution measures. An important precaution measure is having an integrated pilot plant, as a downscaled version of the commercial scale process design. The reasons for needing a pilot plant derived by IPA from 12,000 industrial commercial scale start-up project data [1] are described here.

An integrated pilot plant is needed if either:

A)   The process contains more than 4 new process steps.
B)   The process contains one new process step, and a complex recycle stream over more than 1 process step.
C)   The process contains a novel solids process step.
D)   The process has a crude solids feed stream.

A process step is defined as new if that process step has not been applied before at commercial scale for that specified feed application.

The most common mistake made in process scale-up is that applying conventional unit operations for new applications is that the unit operation is not considered as new, while in reality, it is new.

The background to the criteria, A, B, D is that in processes process steps are connected by mass flows. Because of this integrated nature of processes, by-products in one process step can affect other process steps. If the process contains a recycle stream, then contaminants from the feed and from reactions anywhere in the process can build up high concentrations. This build up can be so slow that it is not noticed in the pilot plant running for a few months, in the commercial scale process this slow build up can in the end cause corrosion, fouling, foaming, or even polymer formation.

The background to criterion C is that solids flow behavior is very different from fluid flow behavior and cannot be treated in design with dimensionless number correlations. Often solids are sticky. A pilot plant will reveal these solids flow problems, so that adequate counteracting measures can be taken.

Scale-up of an integrated process is far from easy. But by the proven method of pilot plant design that is a scaled-down version of the commercial scale design, with long duration tests and construction material samples at critical parts of the process, a successful scale-up is feasible.

https://doi.org/10.1515/9783111203256-030

Scale-up of integrated processes is different from scale-up of individual process steps, such as the scale-up of unit operations. This latter subject is treated in Chapter 25.

24.2 Process scale-up by design method

24.2.1 Process design for no pilot plant needed

The IPA criteria for needing a pilot plant can also be used to design a process in such a way that it does not need an integrated pilot plant. This means that the process should be designed with criteria A, B, C, and D: This a process with:
A) Four, or less new process steps;
B) No complex recycle stream;
C) No novel solids process step;
D) Without a crude solids feed stream.

Ad A: A process step is only not new if it has been applied before for that stream at commercial scale.

Ad B: A recycle stream is only not complex if it is a recycle over the same process step, so it does not include another process step.

Ad C: A solids process step is not new if it has been applied before for that stream at commercial scale.

Ad D: Without a crude solids steam means that none of the feed streams is a crude solids feed stream.

24.2.2 Process design and pilot plant design

There are two ways of scaling up to a commercial scale process implementation. One way is to design and operate a pilot plant and then design the commercial scale process as a Chinese copy of the pilot plant. The other way is first design the commercial scale process and then design the pilot plant as a Chinese copy, just at a smaller scale of the commercial scale design.

The first sequence has the disadvantage that there is no freedom for the designer for the commercial scale design for the selection of equipment types. If the equipment selected for the pilot plant cannot be scaled up to the commercial scale, then multiple pieces of the equipment have to be designed for the commercial scale process.

The second sequence has the advantage that a) only equipment will be chosen that are feasible for the commercial scale capacity and b) if a process design option has a large risk item, then the design may be modified by selecting a process item with a lower risk.

The second has the disadvantage that no pilot plant information is available for the commercial scale design. Hence, this commercial scale design can only be made based on assumptions that the pilot plant will show that the design is feasible. If the pilot plant shows that the design is not feasible, then the commercial scale design has to be changed as a copy of the successfully changed pilot plant.

Vogel of BASF advocates the second sequence with a pilot plant design as a down-scaled version of the commercial scale design [2]. BASF is the most innovative and one of the largest chemical company for over a century, so the second sequence probably proved to be the most successful in BASF's history.

24.3 Example case: Methyl acetate reactive distillation scale-up

Agreda describes the process development by Eastman Chemical Co. of their methyl acetate reactive distillation process in quite some detail [3]. Here, we summarize the main scale-up steps from idea to commercial scale start-up, including various intermediate failures and innovation setbacks. Agreeda was at that time the supervisor of the Advanced Process Technology Group of Eastman Co.

Methyl acetate is formed by the reaction of acetic acid with methanol:

$$\text{Methanol} + \text{acetic acid} = \text{methyl acetate plus water}$$

This is an equilibrium reaction. So, single pass complete conversion is not possible.

Moreover, methyl acetate forms a minimum boiling azeotrope and so does methyl acetate and water.

A process design based on conventional distillations and extractions required 11 process unit operations and several recycle streams. Reactive distillation seemed to be a good candidate for reducing the number of unit operations. However, early attempts in 1921 by Backhaus failed because azeotrope formation was not recognized [3]. Later attempts with reaction distillation were more successful, however, still a second distillation column to recover the product methyl acetate methanol azeotrope was still needed. Impurities in the feeds to the process caused additional problems of intermediate boiling mixtures that contaminate the product and cause build-up in recycle streams [3].

For these reasons, Eastman designed in 1980 a conventional process with two reactors and eight distillation columns. Pilot plant work for this process design was also completed.

However, the desire to have a far better design based on reactive distillation was still pursued in parallel. Eastman process engineers have followed a course on functional process design including options for reactive distillation and used this information to design a single column reactive distillation [4]. The column integrated seven tasks using vapor flows and liquid flows for stripping and extraction. A computer model showed that a single column design could work [3]. Table 24.1 contains all experimental set-ups [3].

Table 24.1: Experimental development set-ups for the methyl acetate reactive distillation process.

Set-up	Details design	Purpose
Laboratory	Not reported	Kinetics, catalyst, equilibria
Lab column	Sieve trays and packings; Capacity 0.05 kg/h	Proof of concept, liquid hold-up optimized
Bench-scale glass column	D = 0.1 m, H = 9 m Capacity: 2 kg/h	Composition profiles as function of conditions Impurities side draw-off, corrosion tests Process control
Pilot plant	D = 0.20 m, H = 30 m Capacity: 45 kg/h Downscaled from commercial scale design	Feasibility and controllability commercial scale Demonstration product purity intermediate side draw-off Corrosion tests

The whole development from concept to the start of detailed engineering of commercial scale design was achieved in less than 1 year. Experimental columns and process modeling were executed partly concurrently. The commercial scale design contained one reactive distillation column followed by three distillation columns. The distillation column for methyl recovery appeared in general not to be needed.

The commercial process started in May 1983. The start-up time to steady state operations of the commercial scale process is not reported, but the commercial scale was highly successful [3].

This highly innovative reactive distillation process development followed the critical success factor of having a downscaled pilot plant with all critical items of the commercial scale plant design.

Bibliography

[1] Harmsen J. Industrial Process Scale-Up: A Practical Innovation Guide from Idea to Commercial Implementation. Elsevier; 2019.

[2] Vogel GH. Process Development: From the Initial Idea to the Chemical Production Plant. John Wiley & Sons; 2005 May 6.

[3] Agreda VH, Heise WH. High-purity methyl acetate via reactive distillation. Chemical Engineering Progress. 1990;86(2):40–6.

[4] Harmsen J, Verkerk M. Process Intensification: Breakthrough in Design, Industrial Innovation Practices, and Education. Walter de Gruyter GmbH & Co KG; 2020 Jul 20.

25 Unit operation scale-up

25.1 Basics Unit operation scale-up by design

Unit operations are the design building blocks of process designs. Their performances are affected by four major performance phenomena:
- Residence time distribution
- Mass transfer distribution
- Heat transfer distribution
- Impulse transfer distribution

These performance phenomena are in general scale dependent. This means that scale-up to commercial scale needs specific attention. Therefore, specific scale-up methods have been developed for unit operations. These are:
- Numbering up
- Dimensionless number analysis
- Dimensionless number correlations
- Modeling
- Empirical step-by-step
- Hybrid empirical modeling

Ad Numbering up. This scale-up method involves having for the commercial scale many parallel units of the same size, flow rates, and conditions as the experimental test unit. The only critical scale-up aspect is the feed distribution over all parallel units. The method is applied for instance for multi-tubular reactors and membranes [1].

Ad Dimensionless number analysis. This method is called the Buckingham theorem scale-up. It involves writing down on physical and chemical phenomena with all their dimensions. Then derive all dimensionless numbers determining the phenomena and then at scale-up from a test unit keeping all these dimensionless numbers at their same value. A full description is provided by Zlokarnik [2].

No description is found for industrial scale applications of this method. There may be two reasons why this method is not applied in industry. First of all, it is hard to keep all dimensionless numbers at the same value for the commercial scale as the test unit. Second, it is hard to explain to nonchemical engineers that this is a reliable method.

The method itself, however, is the foundation of the dimensionless number correlation method described below.

Ad Dimensionless number correlation. This scale-up method is based on dimensionless numbers for performance phenomena such as residence time distribution, mass transfer, and heat transfer. For many unit operations, established dimensionless number correlations are available. For a specific scale-up case, a small scale experimental facility can be applied to validate the correlation experimentally.

https://doi.org/10.1515/9783111203256-031

This method is applied in the process industry for all kinds of unit operations. Correlations can be found in chemical engineering handbooks such as Perry [3].

Ad modeling. Modelling for scale-up of process steps is a big subject and modeling knowledge and modeling computer tools are being developed all the time. In this book, we limit this to modeling of reactors by Computational Fluid Dynamics (CFD), by which the reactor performance is simulated for both reactor configurations and for reactor conditions such as feed concentrations, temperature, and pressure.

It is important to stress that these models have to be validated by experiments, although the model descriptions may state that the modeling tool is based on first principles. For CFD modeling, however, turbulent flow assumptions have to be made on the turbulent field description.

The CFD method is applied in industry for reactors with complex configurations. An elaborate example is described for a crossflow horizontal bubble column with baffles. The baffle structure has been optimized by CFD modeling of the gas and liquid flow, gas-liquid mass transfer, and fast reactions in the liquid [4].

Ad Empirical step-by-step modeling. This scale-up method involves having experimental process units at increasing scale. At each scale, the recipe and the conditions are adjusted to obtain the same product quality. The adjustments at each scale may be correlated with the scale to determine a trend to limit the size of the empirical program.

Ad Hybrid empirical modeling. This scale-up method uses a model to reduce the number of empirical steps by predicting the scale-up effects.

Table 25.1 provides an overview of unit operations and their scale-up methods. It shows that for each unit operation at least one scale-up method is available. Details of scale-up method applications are provided in Section 25.2.

Table 25.1: Unit operations and their applied scale-up methods.

Unit operation	Numbering up	Dim. num. correlations	CFD modelling	Empirical	Hybrid
Reaction Eng. single phase tube	+	+			
Reaction Eng. Packed bed	+	+			
Reaction Eng. Fluid bed			+		+
Reaction Eng. 2/3-ph Bubble Col.			+		+
Distillation		+			
Absorption		+	+		
L-L extraction packed bed		+			
Adsorption	+	+			
Membrane	+				
Static mixer	+	+			
Mechanically stirred tank reactor/mixer			+	+	+

25.2 Unit operations scale-up methods

25.2.1 Scale-up by numbering up

Unit Operations scale-up by numbering up means that the commercial scale unit operation is a set of multiple parallel unit operations, all the same as the individual test unit of the development stage. Not only their size but also their velocities are the same as in this test unit. An example is a commercial scale multitubular fixed bed design, while the pilot plant is a single tube. The only scale-up risk is maldistribution of gas and or liquid over each device of the commercial scale design.

This numbering up also means that for the design of the pilot plant a good sizing of the commercial scale process is needed, so that the pilot plant is a scaled-down version of the commercial scale process design. Table 25.1 shows unit operations suitable for numbering up.

25.2.2 Scale-up by correlations

Unit operations scale-up by dimensionless number correlations is the most common practice in process industries. For many unit operations, these correlations are available in handbooks of chemical engineering; such as Perry's Chemical Engineers' Handbook [3].

It is essential to check the experimental base and the validity range of the dimensionless number correlations. Often the correlations have been determined by a single fluid, often water, and also often for limited geometry dimensions.

If the unit operation is a new process step, according to the IPA definition, meaning that it has not been applied before for that particular feed stream, then experimental validation of the design is needed.

This validation can be obtained by a mini-plant set-up; a downscaled version of the commercial scale unit operation. The experimental set-up should be in the validity range of the correlation. So, if the correlation is valid for the turbulent Reynolds regime, then the experimental set-up should also be operating in that turbulent regime.

25.2.3 Scale-up by CFD modeling

Unit operations scale-up by CFD modeling is applied more and more, now that CFD packages are applicable in many cases. CFD modeling is particularly useful, if the geometrical configuration of the unit operation is complex and the phenomena are complex. An example of such a complex configuration is the formation of an organic peroxide in a horizontal bubble column reactor with baffles, in which mass transfer of oxygen in air to the liquid takes place and the reaction is fast [4]. By CFD modeling, the baffle and air distributor could be optimized such at dissolved oxygen is present everywhere in the liquid. The CFD model was validated experimentally. A full description is available [4].

25.2.4 Empirical step-by-step scale-up

Empirical scale-up means that experiments are carried out step by step at ever increasing scale until the commercial scale operation is reached. The number of steps applied varies between four and five.

Step 1: Experiments are carried out at a small scale and parameters such as temperature, Ph, concentrations, and fluid flow conditions, such as stirrer speed, are varied to obtain the desired performance. Trends in performance related to the parameters are recorded.

Step 2: Experiments at a larger scale, then step 1: typically, a scale-up factor of 10–50 is applied, is carried out, and parameters are varied to reach again to obtain the desired performance. The effects of the larger scale are recorded.

Step 3: Experiments at a larger scale, then step 2 is carried out and parameters are varied again to obtain the desired performance.

Step 4: Experiments at a larger scale, then step 3 is carried out. This step can be the commercial scale plant. Parameters are varied to obtain the desired performance. Trends observed in the scale-up steps 1–3 are used to limit the number of experiments.

This empirical scale-up method is mainly applied in pharma and fine chemicals. The scale-up steps are mostly carried out in existing pilot plants with mechanically stirred tank reactors, operated batchwise.

It sounds like an expensive scale-up method. However, in reality, pharma companies have all pilot units of all scales already in place. For each new pharma product, they only have to perform experimental tests. So, the main costs are the operation cost and not capital investment cost.

25.2.5 Hybrid empirical-modelling scale-up

In hybrid empirical-modeling, scale-up CFD modeling is applied to reduce the number of scale-up steps. The CFD predictions are first validated at step 1. If the CFD predictions are not in agreement with the experimental results and if the predicted trend of variables, such as stirred speed, are not in agreement with the experimental results, then the CFD model will be adjusted.

If the CFD model predictions are in agreement with the experimental results, then one or more scale-up step will be omitted. A few experiments at an intermediate scale will be carried out and compared with the CFD model predictions. If the predictions are in agreement with the experimental results, then the CFD model will be used to determine the commercial scale conditions to obtain the desired performance.

This hybrid modeling is, for instance, applied in biotech industries for large scale fermentations in mechanically stirred tanks. The CFD modeling involves gas-liquid mass transfer and ingredient mixing in the liquid.

25.2.6 Unit Operations selection in view scale-up

Unit operations selection in view of their scale-up reliability happens often for multiphase reactors. A classic case is that of reactors for Fischer–Tropsch synthesis of hydrocarbons from syngas (carbon monoxide and hydrogen). Three-phase bubble column reactors have been applied at commercial scale by Sasol. Shell [5] has applied three-phase trickle flow packed bed reactors.

Shell developed first a three-phase slurry reactor (a sparged bubble column with suspended catalyst particles), with a pilot plant of 1–2 bbl/day production capacity. The investment costs were 10 M$. It started in 1982. In the years to follow, two considerations led to the decision to change from the slurry synthesis reactor to a fixed bed trickle flow reactor. The first consideration was that the scale-up of the slurry reactor, due to uncertainties about scale-up effects, required a costly test plant of 100 bbl/d. The second consideration was that the catalyst development had led to a more stable catalyst, no longer requiring catalyst replacement during operation. Because of these two considerations, a multi-tubular fixed bed synthesis reactor was chosen. This reactor type was applied at commercial scale in Bintulu and in Qatar [5]. Hence, scale-up considerations affected the reactor type choice.

Table 25.1 shows several unit operations and their suitable scale-up methods. For most unit operations, performance correlations based on dimensionless numbers are available, which makes scale-up low cost, by just needing a small-scale test unit to validate the design.

However, for 2- and 3-phase reactors in fluid beds, in bubble columns, and in mechanically stirred tanks, scale-up requires CFD modeling and extensive experimental validation at large scales. So, for these reactor types of development costs are considerable. These costs have to be taken into account in the total cost evaluation in the feasibility stage. The evaluation may result in a different unit operation selection in view of reduced scale-up effort and risks.

25.3 Example case: Hydro-desulfurization reactor scale-up

25.3.1 Case introduction

This case is about the unit operation scale-up selection and applications for a Hydro-DeSulfurization (HDS) reaction of an oil refinery stream. Moonen describes lab-scale experiments for this reaction in a specially designed single string catalyst laboratory test facility and in small-scale fixed bed at trickle flow conditions [7].

Let us treat here a scale-up and scale-down case of the same reaction to illustrate how such a scale-up and scale-down can be executed.

The oil feed fraction contains organic sulfur component, of which the sulfur has to be removed. This can be done by hydrogenation. The HDS hydrogenation reactions involved can be lumped into a reaction stoichiometry:

$$H_2 \text{ (g)} + \text{R-SH (l)} = \text{RH (l)} + H_2S \text{ (g)}$$

The hydrogenation reactions are catalyzed by a heterogeneous catalyst. The porous catalyst particles are so-called extrudates. These are cylindrical particles with diameter 0.7 mm and a length of 3 mm.

All individual reactions are first order in R-S-H components. But because some components react faster than other components, the overall lumped kinetics are higher order in R-SH. Typically, the lumped kinetics are second order in hydrogen.

The hydrogen gas dissolves in the hydrocarbon liquid and the diffuses inside the porous catalyst, where it reacts with the organic sulfur components.

The critical performance phenomena for a reactor are:
- Residence time distribution of the liquid phase in view of deep desulfurization.
- External mass transfer of hydrogen from gas to liquid and from liquid to catalyst and internal catalyst mass transfer

Because the conversion of R-SH is sensitive to the local concentration of hydrogen and R-SH the residence time distribution of the gas and liquid flow is critical to the conversion of R-SH. Because the reactor operates at 150 bar, it is expensive; hence, its volume should be kept minimal this means that all critical performance phenomena are relevant to the reactor design.

The reactor type selection is treated in Chapter 11. The selected reactor type is a cocurrent downflow of gas and liquid over the catalyst particles. This is a so-called trickle bed reactor.

The reactors are designed as adiabatic reactors by feeding a surplus of hydrogen to the reactor. Because hydrogen gas has a high specific heat capacity this is a feasible solution. The hydrogen surplus relative to what is needed for the reaction is typically a factor 10. This means that the hydrogen concentration in the gas phase is nearly uniform; hence, the residence time distribution of the gas phase is not a critical performance phenomenon.

Trickle-bed reactor scale-up

The suitable scale-up method for this packed bed reactor is according to Table 25.1, the dimensionless number correlation scale-up method. So, correlations have to be found for the residence time distribution of the liquid. Potential mass transfer limitations of hydrogen from the gas to the pores inside the catalyst should be evaluated

25.3.2 Commercial scale design

The commercial scale HDS trickle-bed reactor concept design is based on kinetics of a new catalyst to be used. The kinetics have been determined for the new catalyst in a specially designed spinning disk reactor set-up. A check with the Thiele modulus reveals that internal mass transfer limitation is absent for the experimental range.

First, a simple concept design is made for a desired conversion of organic sulfur components of 95 %, (X = 0.95) assuming plug flow for gas and liquid flows and no mass transfer limitations for hydrogen from gas to liquid and from liquid to the porous particles. This results in a concept design with a reactor diameter of 3 m, and a bed height of 7 m. The Liquid Hourly Space Velocity (LHSV) is $7\,h^{-1}$. The superficial liquid velocity is $2\,10^{-3}$ m/s. The Reynolds number is Re = 1.

The hydrogen flow is chosen such that the adiabatic temperature rise of gas plus liquid flow 20 °K. This results in a superficial gas velocity of 0.2 m/s.

The porous catalyst extrudate particles have a characteristic hydraulic particle diameter of 0.001 m. Then this design is checked for all relevant critical performance phenomena.

The Thiele modulus calculation for the first-order reaction rate results in $\phi = 0.2$.

Ad initial gas and liquid distribution
For the uniform distribution of liquid and gas on the top of the catalytic bed, a patented special distributor design will be applied in the detailed engineering part of the Engineering, Procurement, Construction (EPC) stage.

Ad residence time distribution of the liquid phase
For a limited deviation in conversion, the minimum Peclet number is to be calculated from expression (8.12) of Chapter 8.

$$Pe_{min} = 20\,n\,\ln(1/(1 - X)$$

The reaction order for the sulfur components n = 1 and the desired R-SH conversion X = 0.95,

$$Pe_{min} = 24$$

The actual Peclet number value is calculated from

$$Pe = Bo(L/D_p)$$

The Bodenstein number Bo for the liquid flow is provided by Gierman [6]. The liquid flow Reynolds number value of this commercial scale Re = 1.

The Bodenstein number value is then according to Gierman Bo = 0.1.

The Peclet number is Bo $\times$ L/D$_p$ = 0.1 $\times$ 7,000 = 700.

So, the actual Peclet number is much higher than Pe$_{min}$.

So, the commercial scale reactor can be considered to behave as a plug flow reactor.

25.3.3 Laboratory scaled-down design

To confirm the commercial scale concept design of Section 25.3.2, a small-scale trickle bed reactor is designed, which should confirm the conversion for the same catalyst and conditions as the commercial scale. The design is summarized in Table 25.2.

Table 25.2: Down-scaled trickle bed HDS case design.

Design parameter	Dimension	Downscaled	Commercial scale
Conversion R-SH	–	0.95	0.95
Diameter catalyst particle	10^{-3} m	1	1
Diameter inert dilution particle	10^{-3} m	0.2	–
Diluting particle fraction	–	0.2	–
Effective particle diameter	10^{-3} m	0.6	1
Bed diameter	10^{-3} m	20	3,000
Bed length	m	1	7
Inert top layer height	10^{-3} m	100	
Pe$_{min}$	–	24	24
Pe	–	24	700
LHSV	h^{-1}	7	7
Liquid superficial velocity	10^{-3} m/s		2

The first problem to solve is the heat loss through the wall a small-scale reactor, which has a far larger outer surface area per unit reactor volume than the commercial scale reactor. This problem is in general solved by placing insulation material around the reactor. Sometimes controlled electrical heating at the outer reactor wall is applied with a control system to keep the wall temperature the same as the reactor temperature.

For the further scaled-down design, Gierman provides following design rules [6].

Catalyst sample fraction

Because the catalyst amount on the small scale is only a few grams, it is important that the catalyst sample is representative for the whole catalyst batch made for the laboratory test. For a catalyst activity standard deviation of 5 % the catalyst activity, the catalyst sample should be at least 10 % of the catalyst batch made.

Catalyst loading

The catalyst particles are diluted by inert fine Silica Carbide (SiC) particles of 0.2 mm to obtain more easily the required narrow residence distribution of the liquid flow.

To get a uniform local loading of both the catalyst particles and the silica carbide particles, small portions of catalyst and 2SiC are alternately loaded.

Initial liquid feed distribution

The liquid is uniformly distributed over the cross-section of the catalyst bed by placing an inert layer of particles of the same size as the catalyst particles on the top of the bed. The inert height should be five times the bed diameter.

Reactor diameter to avoid liquid wall flow

To avoid wall flow of liquid, the reactor inner diameter is designed as $D_{reactor} >$ $25\,D_{particle}$.

The available lab-scale reactor diameter = 2 cm and the catalyst particle is 0.1 cm. So, design rule is not fulfilled. Gierman advises then to use small inert Silica carbide particles of 0.2 mm to reduce the effective particle size to avoid preferential wall flow. The average particle size should then be 0.8 or smaller.

By taking 20 % of this inert bed dilution material, the effective particle size calculated by the Sauter mean particle size expression, becomes 0.6 mm.

This results in $D_{reactor} > 30\,D_{particle}$.

Bed length to achieve plug flow for the liquid flow

The liquid residence time distribution to achieve plug flow is obtained for Peclet number > 24, according to Section 25.3.2.

The liquid residence time distribution for trickle flow at extremely low Reynolds number is according to Gierman given by $Bo = 2\,10^{-2}$.

The Peclet number is L/D_p . Bo.

This means that L/D_p should be at least 1,200. This results in a bed length of 1 m.

The superficial liquid velocity for the same residence time (same LHSV) is then factor 7 lower than the commercial scale. So, $v_{sup} = 0.3$ mm/s.

Liquid wetting of catalyst particles

Gierman then states that each particle should be wet by the flowing liquid. Gierman provides for this criterion a wetting number W,

$$W = \eta_l v_{sup}/(D_p^2 \rho_l g)$$

for complete wetting of hydrocarbons on porous catalyst particles $W > 5\,10^{-6}$.

By filling in the properties of the liquid results in W = 500 10^{-6}.
So, the particles are fully wet.

Liquid catalyst mass transfer

The liquid side mass transfer at the catalyst particle surface is not the limiting factor as long as mass diffusion inside the reactor is diffusion limited. This is the case for this catalyst and its conditions.

Details of mass transfer limitations outside and inside porous catalysts are treated in Section 10.2 of this book.

Bibliography

[1] Harmsen J. Industrial Process Scale-Up: A Practical Innovation Guide from Idea to Commercial Implementation. Elsevier; 2019 May 16.
[2] Zlokarnik M. Scale-Up in Chemical Engineering. John Wiley & Sons; 2006 Aug 21.
[3] Green DW. Perry's Chemical Engineers' Handbook. 9th ed. New York: McGraw Hill; 2016.
[4] Harmsen J, Bos R. Multiphase Reactors: Reaction Engineering Concepts, Selection, and Industrial Applications. De Gruyter; 2023 Apr 27.
[5] van Helvoort T, van Veen R, Senden M. Gas to Liquids—Historical Development of GTL Technology in Shell. Amsterdam: Shell Global Solutions; 2014.
[6] Gierman H. Design of laboratory hydrotreating reactors: scaling down of trickle-flow reactors. Applied Catalysis. 1988 Jan 1;43(2):277–86.
[7] Moonen R, Alles J, Ras EJ, Harvey C, Moulijn JA. Performance testing of hydrodesulfurization catalysts using a single-pellet-string reactor. Chemical Engineering & Technology. 2017 Nov;40(11):2025–34.

26 Vocabulary

26.1 Basic vocabulary

The chemical engineering discipline has its own terms with specified definitions. The Oxford Dictionary of Chemical Engineering [1] is a particularly good book to find definitions for most chemical engineering terms and words. Here, terms are placed, which need more background information, and terms, which are sometimes misused.

The terms are grouped in subject sections so that the relations between the terms for each subject are clear.

26.2 Innovation stage-gate model terms

New process industry branches with innovative processes are now emerging: Green Hydrogen Production, Waste-to-Value, Renewable Chemicals for Textiles, and Sustainable Packaging Materials. Start-up companies often develop their innovative processes. These companies have enormously motivated and creative personnel, by which they solve all kinds of technical and financial problems.

However, in some cases, I have noted that available knowledge on innovation methods and industrial process scale up practices are not known to these start-ups companies, by which process development is hampered and commercial scale start-up fails.

Here are search terms for available knowledge in the public domain on innovation models to avoid this.

In the *Discovery stage*, new ideas are generated and proof of principle experiments are carried out. This stage is also called *Ideation stage*.

In the *Concept stage*, ideas are defined by concept design and proof of concept experiments are carried out.

In the *Feasibility stage*, the commercial feasibility of the new product and or process is defined. In addition, the development plan is made. The overall economics of the commercial plant and the development cost are determined leading to the decision to continue or stop the project.

The Valey of Death is the costly development stage, where the project sometimes stops.

In the *Development stage*, the new product and or process is tested in a mini-plant or pilot plant. The stage ends with the *Front-End-Engineering Design (FEED)*.

Front-End-Engineering Design (FEED) is a collection of documents, i. e., a *Front-End Design* (also called a *Design Engineering Package*), investment cost, and economics a hazard and operability description, and a risk document.

The *Engineering-Procurement-Construction (EPC) stage* results in the commercial scale process.

https://doi.org/10.1515/9783111203256-032

In the *Start-up stage* starting with the commissioning of the new plant, the new product and or new process is produced for market sales.

The terms are described in some detail by Harmsen [2].

26.3 Technology Readiness Levels terms

Technology Readiness Levels (TRL) define the degree of readiness of a new technology. It has nine levels from experimental proof of principle to 2nd commercial scale operation. It helps in assessing new technologies to be purchased, because it shows its present status and so what still needs to be done to commercialize it. The levels state what has been done to reach that level. TRL has been developed by NASA, so that any new piece of technology proposed by a technology provider or university could be quickly assessed. The general descriptions of the levels are therefore very abstract [3]. Here, the TRL levels are described for new processes. I added TRL 0, so that also just an idea before experimental work has been done, and can also be given a level [2].

TRL 0: The technology idea is stated
TRL 1: Experimental proof of principle individual key new elements
TRL 2: A process concept design is available
TRL 3: Feasibility proof of process concept by a technoeconomic assessment
TRL 4: Process concept is experimentally validated by integrated mini-plant
TRL 5: Industry professionals assess process technoeconomics
TRL 6: Process is demonstrated by pilot plant in a process industry
TRL 7: First commercial scale plant is in steady state operation
TRL 8: Learnings from first commercial scale plant are implemented in new design
TRL 9: Second commercial scale plant is in steady-state operation.

TRL values are often loosely used. To increase the reliability of TRL values, people qualified to determine certain TRL are stated here.

TRL 0–4 can be determined by research and developers in academia and industry.

TRL 5–6 are about technologies applied in an industrial environment. Industrial practitioners can therefore only determine these.

TRL 7–9 are about technologies applied in a commercial environment. These TRLs can only be determined by commercial scale practitioners.

26.4 Scale-up methods

Scale up methods are found in various books. Harmsen made a summary of the methods [2]. Here follow definitions of the methods.

Integral Processes consist of several unit operations connected to each other, transforming input streams to product streams.

Integral process scale-up is a method for a new integral process consisting of several unit operations. The method describes all innovation stages to commercial scale start-up. The innovation stages are ideation, concept, feasibility, development, Engineering Procurement Construction, Start-up.

Unit Operation scale-up is a method for individual unit operations. For this scale-up, the following scale-up methods are available.

Correlation based scale-up is the most common scale-up method for established unit operations. It is based on dimensionless number correlations for specific performance phenomena, residence time distribution, mixing, mass transfer, and heat transfer. These correlations are used to make commercial scale unit operation designs. In a downscaled test, unit applied correlations can be validated.

Computational Fluid Dynamics (CFD) model-based scale-up is about the use of CFD model to design and optimize equipment shapes and internals. It is used for unconventional process steps. The CFD model can include reactions, multiphase fluid flow, mixing, mass transfer, and heat transfer phenomena. It facilitates process equipment configuration optimization. Experimental validation of most critical model items can be executed in cold-flow models, hot pilot plants, and also in commercial scale equipment using fluid flow measuring probes.

Numbering up scale-up is about having many of the same unit operations in parallel of the commercial scale process. One or a few of these unit operations may have been tested in a pilot plant. This method is also called brute force scale-up, linear scale-up, scale-out, and German scale-up.

Empirical scale-up is a step-by-step scale-up method of experimentation on increasing scales from laboratory bench scale, micro-plant, mini-plant, small pilot plant, large pilot plant, demo plant, commercial scale plant. In each step process, parameters are used to optimize the performance.

Critical performance phenomena of individual process unit operations, such as residence time distribution, mass transfer, and heat transfer help to select design methods and select a scale-up method.

26.5 Process design

The field of process concept design with the discipline of chemical engineering is still expanding and new terms appear. Here is a brief description of these novel terms entered into the vocabulary.

Process concept design is a systematic procedure to find the best flowsheet, i. e., to select the best process units and the interconnecting streams among these units and estimate the optimum design conditions [Douglas]. With process units, Douglas means conventional unit operations.

The term *flowsheet* means the representation of the process design by process units and their connecting streams [4, 5]. The term is also used for computer programs by which the connecting stream compositions are calculated based on mathematical models of the process units.

Process synthesis is also a term used for process concept design, to generate the best flowsheet structure to convert feedstock into product [1]. However, process synthesis is also used for process concept design, not based on conventional unit operations, but based on *process functions* (also called tasks) and streams [6]. The functions are simple transformations, such as reaction, mixing, and enthalpy change, describing what transformation is needed, but not how. In a second step, the functions are further defined for physical and chemical conditions. In a third design step, functions and also streams are integrated as much as is optimal. For the other functions, suitable unit operations are selected. Finally, a flow sheet is generated, which can contain a novel process step, such as a reactive, extractive, distillation column combining several functions [6].

Functional process step is a term introduced by Bridgewater [7] to simplify process design representations. A functional process step contains all unit operations connected to each other with the same function. A distillation train for example forms a single functional process step. Also, several reactors in series or in parallel form a single functional process step, and helping in making process concept designs revealing all functions needed and all essential process streams.

Unit operations is a process step characterized by the same physical and chemical rules as for any application. The term was introduced by AD Little in 1916. Perry's Chemical Engineers' Handbook describes most of them [8]. By process intensification, research novel unit operations are generated based on novel dynamic operations, and different external fields, such as gravity, microwaves, and light [9].

26.6 Process intensification

Process intensification is a set of radically innovative process design principles, which can bring significant benefits in terms of efficiency, cost, product quality, safety, and health [9], over conventional process designs based on unit operations [10].

Modular design is a detailed process engineering design of modules, which can be easily connected to each other (plug in principle). The modules are detailed process engineered and constructed by an Engineering Procurement Construction (EPC) company [11]. Each module can be constructed many times using the same detailed design [11].

Skid mounted means that the process plant or modules are constructed inside a skid frame. The process plant is constructed at the EPC company, dry tested, and then transported to the manufacturing site [11]. Most modular designed processes are skid mounted.

Distributed process plants means that many small capacity process plants are distributed over various locations.

To keep the investment cost per ton of products at acceptable levels, often the process concept design is based on process intensification methods, while the detailed process engineering is based on modular design and the construction is skid mounted.

Stick built is the conventional constructing of process plants at the manufacturing site. It is the opposite of modular design with skid mounted construction.

Building blocks are the process intensified versions of unit operations of a process design [11].

Functional process design is conceptual process design based on functions and function integration [10, 11] rather than on established unit operations.

Numbering up is a term used by many people and often each means something different. The presenter may mean having many of the same equipment pieces parallel to each other but connected to the same input and output device. The presenter may mean having process trains in parallel to each other on the same manufacturing site. The presenter may mean a linear investment cost curve with production capacity. Often the last meaning is wrongly connected to the other meanings. Even with parallel process trains the total investment cost correlates in most cases with the capacity to a power less than 1. As a lot of construction cost, such as buildings, concrete floors, pillars, and others are still shared. So, always ask: "what do you mean." If it is written down, read the whole document to find out what is meant.

26.7 Process scale-up

Scale-up is generating knowledge to transfer ideas into successful commercial implementations [2]. An alternative definition. Scale-up is a larger production scale [12].

Scale-up factor is the capacity increase from one innovation stage to the next stage [13].

Micro-flow reactor (or process) is a continuously operated reactor (or process) at laboratory scale [2]. An alternative definition, "Microreactor technology is concerned with chemical reactions and unit operations in components and systems whose characteristic dimensions range from submillimeter down to submicro-meter region" [13].

Mini-plant is a down-scaled version of the commercial scale design made in the feasibility stage [2]. An alternative definition Is that a mini plant is a continuous complete laboratory plant containing reactors separators and recycle streams producing up to 100 gram/h [13].

Pilot-plant is a down-scale version of (a section of) the industrial scale process design to experimentally validate the concept design [2, 13, 14].

Integral pilot plant is a downscaled version of the commercial scale process design. Such an integrated pilot plant is essential for commercial scale success if the process contains more than four new process steps, or a recycle stream, or a new solids processing step, or a crude solids feed. This conclusion results from a statistical study by Independent Project Analysis reported in [2].

Demonstration plant is, according to the TRL 7 description, any first version of a new commercial process in a commercial scale processing environment is a demonstration plant.

Demo plant means a larger scale pilot plant, operated still in the development department of the company. So, it is not a demonstration plant.

26.8 Chemical engineering for pharma

A quiet revolution is occurring in the pharma industry, where now chemical engineering methods are introduced in research, design, development, and implementations by many pharmaceutical companies. This quiet revolution is due to the fact that regulatory bodies such as the Food and Drug Administration (FDA) of the USA and the European Medicines Agency (EMA) play an active role in this [15], and that process technology providers develop design and construction plants fitting to the specific pharma needs [16]. Here is a vocabulary of terms used in this new field.

Quantitative System Pharmacology (QSP). Chemical Engineering as applied to Medicine was a one-day forum meeting the author attended at Paris, December 5, 2022, organized by the EFCE. His main conclusion of the 11 presentations from industry and academia is that chemical engineering methods and practices are now researched in many academic groups and are integrated in the industry, to rapidly find new medicines for specific diseases. The companies call this integration *Quantitative System Pharmacology (QSP)*.

The methods applied are:
– *Chemical reaction engineering* concepts and models to generate medicine options
– *Designing, constructing, and testing with physical analogues* using chemical engineering principles such as hydrodynamics, mass transfer, and reaction.
– *Modular flexible design and operation*

Physical analogue can contain an analogue of the human body, such as the lungs, and contain the medicine admission technique, such as an inhaler. With such an experimental set-up, both the medicine, its formulation, and the inhaler can be optimized. This new way of working facilitates rapidly finding new medicines and also reduces animal testing programs.

Modular flexible design and operation means that the plant contains several modules that can be moved out or into production by running campaigns. Typically, 8 to 12 different medicines or Active Pharma Ingredients (API's) are produced per year in this way. This means also that cleaning in between campaigns is an essential process step. This change of modules can be made for some plants in a few days. David Lindner of Hoffman La Roche mentioned that they had combinations of continuous (80 %) and batch (20 %) processing modules [16].

Flow chemistry Flow chemistry means continuous processing (rather than batch processing) for a new medicine or Active Pharma Ingredient *(API)*. Often it means, in addition, which process intensified technologies are used, such as micro flow reactors. The term is mainly used for lab-scale continuous processes.

Scale-up in the pharma industry means nearly always product scale-up. It means that in several phases more of the new active pharmaceutical ingredient or the new medicine is produced for specified pharma phases are *discovery, pre-clinical, clinical phase I, clinical phase II, clinical phase III, approval, and finally manufacturing* [17]. When someone wants to talk about new process development in a pharma environment it is better to avoid for that specific activity the word scale-up, otherwise confusion may start. It is better to use the term new process technology introduction.

Quantitative System Pharmacology (QSP) means that chemical engineering methods and practices are applied in pharmaceutical companies to rapidly find new medicines for specific diseases [18].

The two main methods applied are:

a) Chemical reaction engineering concepts and models to generate medicine options
b) Designing, constructing, and testing with physical analogues, using chemical engineering principles such as hydrodynamics, mass transfer, and reaction.

The physical analogue can contain an analogue of the human body, such as the lungs, and contain the medicine admission technique, such as an inhaler.

26.9 Chemical reaction engineering

Chemical reaction engineering is the proper way of bringing to practice chemical reactions. This definition stems from one of the founding fathers of chemical engineering, Octave Levenspiel [19].

Continuously Stirred Tank Reactor (CSTR)

CSTR is a theoretical concept of a fully back-mixed reactor in continuous operation mode. Octave Levenspiel calls it a Mixed Flow Reactor (MFR) [20].

Westerterp calls this concept the Continuously Ideally Stirred Tank Reactor (CISTR). This stresses the assumption that it is ideally mixed [21].

The CSTR term is often misused for a mechanically stirred tank reactor. If such a reactor is operated batchwise, then it has plug flow behavior, so it is then the opposite of a fully back-mixed reactor.

Plug Flow Reactor (PFR)

This is a theoretical concept of a reactor in which each fluid element has the same residence time. So, the fluid flows as a plug through the reactor, or the fluid elements stay the same time in the reactor. This latter can occur, for instance, in a batch reactor, with all feeds entered at the start of the batch operation [20].

The misconception sometimes is that it is about a real pipe reactor. A real pipe reactor, however, may show a considerably wide residence time distribution of the individual fluid elements.

Reaction rate (r)

Reaction rate is the rate at which a molecular species is formed or disappears by a chemical reaction per unit fluid volume. It is an intensive variable, like temperature, or pressure are intensive variables [22].

The reaction rate depends on the local conditions in the reaction system. It is independent on the type of reactor system itself.

26.10 Sustainable design of processes and products

Here is a shortlist of terms relevant to the sustainable design of products and processes.

"*Sustainable development* is not a fixed state of harmony, but a process of change in which the exploitation of resources, the direction of investments, the orientation of technological development, and institutional change are made consistent with future as well as present needs. Sustainable development is development that meets the needs of the present generation without compromising the ability of future generations to meet their own needs" [23].

Sustainable development goals are the main part of the United Nations Sustainable Development Agenda 2030 [24].

Life-cycle analysis concerns all environmental impacts attributable to the functioning of a product over its life cycle, cradle to grave (abbreviated) [25].

Industrial ecology is the study of the flows of materials and energy in industrial and consumer activities, of the effects of these flows on the environment, and of the influences of economic, political, regulatory, and social factors on the flow, use, and transformation of resources [26].

Industrial symbiosis is a subset of industrial ecology. It is concerned with transferring the biology analogy of the ecosystem to the industrial system [27].

A *circular economy* is one that is restorative and regenerative by design, and which aims to keep products, components, and materials at their highest utility and value at all times, distinguishing between technical and biological cycles [28].

26.11 Mixing

The term "mixing" is used in various ways in the field of chemical engineering. Here are some useful definitions. Common misunderstandings of terms are also described.

Mixing

Mixing 1: The intimate contact of components to obtain one homogenous fluid [1].

Mixing 2: In chemical engineering process concept design "mixing" means a process function, or a process building block, or a unit operation, which performs mixing. Also, the term "mixer" is used with the same meaning as mixing [1].

Mixing is also used for dispersing a phase in a different phase such as dispersing a gas as bubbles into a liquid. In general, this use does not lead to misunderstanding, as it is obvious what is meant.

Micro-mixing

If the mixing is down to the uniform molecular level, then it is called micro-mixing. This term is mainly used in chemical reaction engineering, when fast reactions are involved, and the rate of micro-mixing affects the reactor performance in terms of conversion and selectivity [22].

Micro-mixing is however also relevant in other unit operations such as in mixing (mixers).

Back-mixing

The term "back-mixing" is used in chemical reaction engineering for a continuous flow reactor concept with uniform concentrations inside the reactor. It is about the residence time distribution of the flow through the reactor, in which each fluid element has an equal chance of leaving the reactor or staying inside. This concept reactor type is called Mixed Flow Reactor (MFR) by Levenspiel in his textbook Chemical Reaction Engineering [20]. It is also called Continuously Operated Stirred Tank Reactor (CSTR) [22].

The term has nothing to do with an actual reactor. The liquid phase of a bubble column for instance may, under certain conditions, behave as a CSTR, while the actual reactor does not contain a stirrer. A mechanically stirred tank reactor operated batchwise; however, behaves in residence time distribution theory as a Plug Flow Reactor (PFR), which is in residence time distribution behavior the opposite of a CSTR.

If the reactor is also micro-mixed, then the reactor concept is called Continuously operated ideally Stirred Tank Reactor (CiSTR) [21].

Residence time distribution is relevant in designing reactors and in other unit operations such as distillation, extraction, and mixing.

26.12 Membrane separations

The science field membrane separations have its own terminology, as described by Kronos [29], which is different from the chemical engineering field. Here are some major membrane vocabulary term descriptions of Kronos and their translation into chemical engineering terms.

Membrane filtration

In membrane filtration, a component, or several components are filtered from the feed stream, to stay on the feed side of the membrane. This membrane filtration belongs to the chemical engineering unit operation filtration.

Membrane separation

In membrane separation, the desired pure product selectively diffuses through the membrane, while all other components stays on the feed side of the membrane. This then results in a very pure product stream.

Permeate

The component diffusing through the membrane is called permeate. This is in general the main product stream. The outlet chamber is called the permeate chamber.

Retentate

The mixture of all other components staying on the feed side of the membrane is called retentate. The feed side chamber is called the retentate chamber.

Dead-end membrane separation

In dead-end membrane separation, the retentate side chamber is closed at one end, so that only diffusion and flow through the membrane layer can occur. After some time, the operation has to stop and the material at the retentate side is removed. This is in chemical engineering terms a batch operation.

Tangential membrane separation

In tangential membrane separation, the feed flows tangentially along the membrane surface through the retentate chamber and leaves at the retentate side exit, while the desired pure component diffuses through the membrane layer to the permeate side. So, this is a continuous operation.

26.13 Engineering and construction

Detailed engineering means that every process item is designed in size and construction material, which all process control and instrumentation are defined and that the structure supporting all process equipment and piping is designed.

Procurement means procuring all major process equipment.

Construction means the actual fabrication of the process in which all major equipment is installed and connected with pipes and control systems.

Modular design definition

Modular design means that the process design is split up in modules. Each module is designed and then the resulting design modules are standardized. This means that the design can be used again and again for constructing the same process, without any design effort. It also means that if one module of the design is improved, that the other modules do not need a change.

Prefabrication

Prefabrication means that the process is first constructed at the Engineering Procurement Construction (EPC) site. Also, dry testing to validate the technical integrity and proper function of the instrumentation and control is carried out at the construction site. After that, it deconstructed in pieces, transported to the manufacturing site, and reassembled. Ater that, the process starts up to produce the desired product. So, prefabrication differs from construction at the manufacturing site. The latter is called stick built.

Stick built

Stick built means constructing the process at the manufacturing site where the process will be operated.

Skid mounted and container mounted construction definition

To facilitate transportation, the whole process is constructed inside a container or the process is skid mounted. This means that a support structure is attached to the process. If modular design is the basis for prefabrication, then often each module is on its own skid mounted, facilitating transport and easy reassembly of the module at the manufacturing site.

Distributed modular design combined with prefabrication

When several small capacity processes for the same purpose are installed at different locations by the same technology owner, then that is often called distributed processes or distributed plants. Each process plant may be installed sequentially with some time in between each installation implemented [30–32].

Bibliography

[1]   Schaschke C. A Dictionary of Chemical Engineering. Oxford: OUP; 2014 Jan 9.

[2]   Harmsen J. Industrial Process Scale-Up: A Practical Innovation Guide from Idea to Commercial Implementation. Elsevier; 2019 May 16.

[3]   DOE. Technology Readiness Assessment Guide (DOE G 413.3-4). United States Department of Energy, Office of Management; 2011 Sep 15.

[4]   Douglas JM. Conceptual Design of Chemical Processes. New York: McGraw-Hill; 1988.

[5]   Smith R. Chemical Process: Design and Integration. John Wiley & Sons; 2005.

[6]   Siirola JJ. Industrial applications of chemical process synthesis. In: Advances in Chemical Engineering. Vol. 23. Academic Press; 1996 Jan 1. pp. 1–62.

[7]   Bridgewater AV. The functional unit approach to rapid cost estimation. The Cost Engineer. 1974;13(5).

[8]   Green DW, Southard MZ. Perry's Chemical Engineers' Handbook. McGraw-Hill Education; 2019.

[9]   Stankiewicz A, Van Gerven T, Stefanidis G. The Fundamentals of Process Intensification. John Wiley & Sons; 2019.

[10]   Harmsen J, Verkerk M. Process Intensification. Berlin: De Gruyter; 2020.

[11]   Harmsen J, Verkerk M. A new approach to industrial innovation. Chemical Engineering Progress. 2021 Mar 1;117(3):50–3. 3.

[12]   de Haan AB. Process Technology—An Introduction. Berlin: De Gruyter; 2015.

[13]   Vogel GH. Process Development—From the Initial Idea to the Chemical Production Plant. Weinheim: Wiley-VCH; 2005.

[14]   Harmsen J. Product and Process Design—Driving Innovation. Berlin: De Gruyter; 2018.

[15]   Lee SL. Current FDA perspective for continuous manufacturing. In: 2nd International Symposium on Continuous Manufacturing of Pharmaceuticals. Cambridge, USA. 2016. Sep 26.

[16]   ZETON. Pharma-biotech plants for continuous manufacturing of Active Pharmaceutical Ingredients (API). Sourced 222 Nov. 2022 https://www.zeton.com/industries/pharma-biotech/.

[17]   Kahn A. Phase-appropriate formulation and process design. Pharmaceutical Technology. 2016;40(1):42–5.

[18]   Cesar Pichardo of Astra Zeneca and Roberto Abbiati of Boehringer Ingelheim, presentations, Chemical Engineering as applied to Medicine, EFCE forum meeting, Paris, 5th of December 2022.

[19]   Levenspiel O. The coming-of-age of chemical reaction engineering. Chemical Engineering Science. 1980 Jan 1;35(9):1821–39.

[20]   Levenspiel O. Chemical Reaction Engineering. John Wiley & Sons; 1998.

[21]   Westerterp KR, van Swaaij WP, Beenackers AA, Kramers H. Chemical Reactor Design and Operation. John Wiley & Sons; 1984.

[22]   Harmsen J, Bos R. Multiphase Reactors: Reaction Engineering Concepts, Selection, and Industrial Applications. Berlin: De Gruyter; 2023.

[23]   Brundtland GH. Our Common Future, report World Commission on Environment and Development of the UN; 1987.

[24]   UN. Transforming Our World: The 2030 Agenda for Sustainable Development. Resolution General Assembly UN; 2015.

[25]   Schaschke C. Oxford Dictionary of Chemical Engineering. Oxford, UK: Oxford U. Press; 2014.

[26]   Ayres RU, Ayres LW. A Handbook of Industrial Ecology. Cheltenham, USA: Edward Elgar Publ. Ltd; 2002.

[27]   Chertow MR. Uncovering industrial symbiosis. Journal of Industrial Ecology. 2007;11(1):11–30.

[28]   Ellen MacArthur Foundation. Circular economy scheme. https://www.ellenmacarthurfoundation.org/circular-economy/interactive-diagram.

[29]   Koros WJ, Ma YH, Shimidzu T. Terminology for membranes and membrane processes. Journal of Membrane Science. 1996 Nov 13;120(2):149–59.

[30]   Baldea M, Edgar TF, Stanley BL, Kiss AA. Modular manufacturing processes: Status, challenges, and opportunities. AIChE Journal. 2017 Oct;63(10):4262–72.

[31]   Patience GS, Boffito DC. Distributed production: Scale-up vs experience. Journal of Advanced Manufacturing and Processing. 2020 Apr;2(2):e10039.

[32]   Weber RS, Snowden-Swan LJ. The economics of numbering up a chemical process enterprise. Journal of Advanced Manufacturing and Processing. 2019 Apr;1(1–2):e10011.

Index

https://doi.org/10.1515/9783111203256-033